HUTCHINSON

Guide to
Countries of the
World

BROCKHAMPTON PRESS
LONDON

Copyright © Helicon Publishing Ltd 1993

Helicon Publishing Ltd
42 Hythe Bridge Street
Oxford OX1 2EP

Printed and bound in Great Britain by
Mackays of Chatham Plc,
Chatham, Kent

This edition published 1997 by
Brockhampton Press Ltd
20 Bloomsbury Street
London WC1B 3QA
(*a member of the Hodder Headline PLC Group*)

ISBN 1-86019-571-7

British Cataloguing in Publication Data

A catalogue record for this book is available
from the British Library

Editorial director
Michael Upshall

Project editor
Sara Jenkins-Jones

Text editor
Catherine Thompson

Art editor
Terence Caven

Page make-up
Helen Bird

Production
Tony Ballsdon

Afghanistan
Republic of
(*Jamhuria
Afghanistan*)

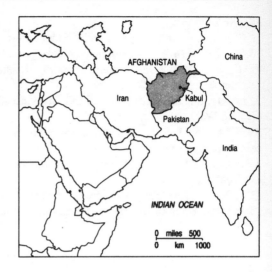

area 652,090 sq km/251,707 sq mi
capital Kabul
towns Kandahar, Herat, Mazar-i-Sharif
physical mountainous in centre and NE, plains in N and SW
environment an estimated 95% of the urban population is without
access to sanitation services
head of state Burhanuddin Rabbani from 1992
head of government to be announced
political system emergent democracy
exports dried fruit, natural gas, fresh fruits, carpets; small amounts of
rare minerals, karakul lamb skins, and Afghan coats
currency afgháni
population (1989) 15,590,000 (more than 5 million became refugees
after 1979); growth rate 0.6% p.a.
life expectancy (1986) men 43, women 41
languages Pushtu, Dari (Persian)
religion Muslim (80% Sunni, 20% Shi'ite)
literacy men 39%, women 8% (1985 est)
GNP $3.3 bn (1985); $275 per head

chronology
1747 Afghanistan became an independent emirate.
1839–42 and 1878–80 Afghan Wars instigated by Britain to counter
the threat to British India from expanding Russian influence in
Afghanistan.
1919 Afghanistan recovered full independence following Third
Afghan War.
1953 Lt-Gen Daud Khan became prime minister and introduced
reform programme.
1963 Daud Khan forced to resign and constitutional monarchy
established.
1973 Monarchy overthrown in coup by Daud Khan.
1978 Daud Khan ousted by Taraki and the Democratic Party of
Afghanistan (PDPA).
1979 Taraki replaced by Hafizullah Amin; Soviet Union entered
country to prop up government; they installed Babrak Karmal in
power. Amin executed.
1986 Replacement of Karmal as leader by Dr Najibullah Ahmadzai.
Partial Soviet troop withdrawal.
1988 New non-Marxist constitution adopted.
1989 Complete withdrawal of Soviet troops; state of emergency
imposed in response to intensification of civil war.
1991 UN peace plan accepted by President Najibullah but rejected by
the mujaheddin. US and Soviet military aid withdrawn. Mujaheddin
began talks with Russians and Kabul government.
1992 April: Najibullah regime overthrown. June: after a succession of
short-term presidents, Burhanuddin Rabbani named interim head of
state; Islamic law introduced. Sept: Hezb-i-Islami barred from
government participation after shell attacks on Kabul. Dec: Rabbani
elected president for two-year term by constituent assembly.
1993 Jan: renewed bombardment of Kabul by Hezb-i-Islami and other
rebel forces. Interim parliament appointed by constituent assembly.
March: peace agreement signed between Rabbani and dissident
mujaheddin leader Gulbuddin Hekmatyar, under which Hekmatyar to
become prime minister.

Albania
Republic of
(*Republika e
Shqipërisë*)

area 28,748 sq km/11,097 sq mi
capital Tiranë
towns Shkodër, Elbasan, Vlorë, Durrës
physical mainly mountainous, with rivers flowing E–W, and a narrow
coastal plain
head of state Sali Berisha from 1992
head of government Alexander Meksi from 1992
political system emergent democracy
exports crude oil, bitumen, chrome, iron ore, nickel, coal, copper wire,
tobacco, fruit, vegetables
currency lek
population (1990 est) 3,270,000; growth rate 1.9% p.a.
life expectancy men 69, women 73
languages Albanian, Greek
religion Muslim 70%, although all religion banned 1967–90
literacy 75% (1986)
GNP $2.8 bn (1986 est); $900 per head

chronology

c. 1468 Albania made part of the Ottoman Empire.

1912 Independence achieved from Turkey.

1925 Republic proclaimed.

1928–39 Monarchy of King Zog.

1939–44 Under Italian and then German rule.

1946 Communist republic proclaimed under the leadership of Enver Hoxha.

1949 Admitted into Comecon.

1961 Break with Khrushchev's USSR.

1967 Albania declared itself the 'first atheist state in the world'.

1978 Break with 'revisionist' China.

1985 Death of Hoxha.

1987 Normal diplomatic relations restored with Canada, Greece, and West Germany.

1988 Attendance of conference of Balkan states for the first time since the 1930s.

1990 One-party system abandoned; first opposition party formed.

1991 April: Party of Labour of Albania (PLA) won first multiparty elections; Ramiz Alia re-elected president; three successive governments formed. PLA renamed Socialist Party of Albania (PSS).

1992 Former communist officials arrested on corruption charges. Presidential elections won by Democratic Party of Albania (DP); Sali Berisha elected president. Alia charged with corruption and abuse of power; totalitarian and communist parties banned.

1993 Jan: Nexhmije Hoxha, widow of Enver Hoxha, sentenced to nine years' imprisonment for misuse of government funds 1985–90.

Algeria
Democratic and
Popular Republic of
(*al-Jumhuriya al-
Jazairiya ad-
Dimuqratiya ash-
Shabiya*)

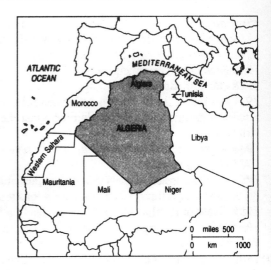

area 2,381,741 sq km/919,352 sq mi
capital al-Jazair (Algiers)
towns Qacentina/Constantine, Ouahran/Oran, Annaba/Bône
physical coastal plains backed by mountains in N; Sahara desert in S
head of state Ali Kafi from 1992
head of government Belnid Absessalem from 1992
political system semi-military rule
exports oil, natural gas, iron, wine, olive oil
currency dinar
population (1990 est) 25,715,000 (83% Arab, 17% Berber); growth
rate 3.0% p.a.
life expectancy men 59, women 62
languages Arabic (official); Berber, French
religion Sunni Muslim (state religion)
literacy men 63%, women 37% (1985 est)
GDP $64.6 bn; $2,796 per head
chronology
1962 Independence achieved from France. Republic declared. Ahmed
Ben Bella elected prime minister.

1963 Ben Bella elected Algeria's first president.
1965 Ben Bella deposed by military, led by Colonel Houari
Boumédienne.
1976 New constitution approved.
1978 Death of Boumédienne.
1979 Benjedid Chadli elected president. Ben Bella released from
house arrest. National Liberation Front (FLN) adopted new party
structure.
1981 Algeria helped secure release of US prisoners in Iran.
1983 Chadli re-elected.
1988 Riots in protest at government policies; 170 killed. Reform
programme introduced. Diplomatic relations with Egypt restored.
1989 Constitutional changes proposed, leading to limited political
pluralism.
1990 Fundamentalist Islamic Salvation Front (FIS) won Algerian
municipal and provincial elections.
1991 Dec: FIS won first round of multiparty elections.
1992 Jan: Chadli resigned; military took control of government;
Mohamed Boudiaf became president; FIS leaders detained. Feb: state
of emergency declared. March: FIS ordered to disband. June: Boudiaf
assassinated; Ali Kafi chosen as new head of state and Belnid
Absessalem as prime minister.

Andorra
Principality of
(*Principat
d'Andorra*)

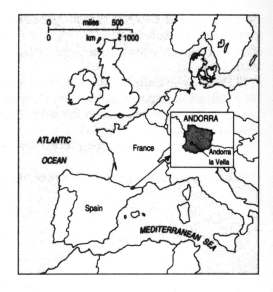

area 468 sq km/181 sq mi
capital Andorra-la-Vella
towns Les Escaldes
physical mountainous, with narrow valleys
heads of state Joan Marti i Alanis (bishop of Urgel, Spain) and
François Mitterrand (president of France)
head of government Oscar Riba Reig from 1989
political system semi-feudal co-principality
exports main industries are tourism and tobacco
currency French franc and Spanish peseta
population (1990) 51,000 (30% Andorrans, 61% Spanish, 6% French)
languages Catalan (official); Spanish, French
religion Roman Catholic
literacy 100% (1987)
GDP $300 million (1985)
chronology
1278 Treaty signed making Spanish bishop and French count joint
rulers of Andorra (the king of France later inherited the count's right).

1970 Extension of franchise to third-generation female and second-generation male Andorrans.
1976 First political organization (Democratic Party of Andorra) formed.
1977 Franchise extended to first-generation Andorrans.
1981 First prime minister appointed by General Council.
1982 With the appointment of an Executive Council, executive and legislative powers were separated.
1991 Andorra's first constitution planned; links with European Community (EC) formalized.
1993 First constitution approved in a referendum.

Angola
People's Republic of
(*República Popular
de Angola*)

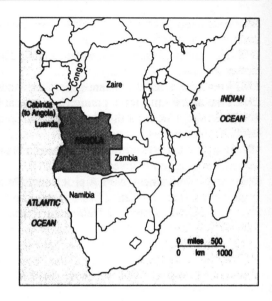

area 1,246,700 sq km/481,226 sq mi
capital and chief port Luanda
towns Lobito and Benguela, Huambo, Lubango
physical narrow coastal plain rises to vast interior plateau with
rainforest in NW; desert in S
head of state and government José Eduardo dos Santos from 1979
political system socialist republic
exports oil, coffee, diamonds, palm oil, sisal, iron ore, fish
currency kwanza
population (1989 est) 9,733,000 (largest ethnic group Ovimbundu);
growth rate 2.5% p.a.
life expectancy men 40, women 44
languages Portuguese (official); Bantu dialects
religions Roman Catholic 68%, Protestant 20%, animist 12%
literacy 20%
GDP $2.7 bn; $432 per head

recent chronology
1951 Angola became an overseas territory of Portugal.
1956 First independence movement formed, the People's Movement for the Liberation of Angola (MPLA).
1961 Unsuccessful independence rebellion.
1962 National Front for the Liberation of Angola (FNLA) formed.
1966 National Union for the Total Independence of Angola (UNITA) formed.
1975 Independence achieved from Portugal. Transitional government formed from representatives of MPLA, FNLA, UNITA, and Portuguese government. MPLA proclaimed People's Republic of Angola under the presidency of Dr Agostinho Neto. FNLA and UNITA proclaimed People's Democratic Republic of Angola.
1976 MPLA gained control of most of the country. South African troops withdrawn, but Cuban units remained.
1977 MPLA restructured to become the People's Movement for the Liberation of Angola–Workers' Party (MPLA–PT).
1979 Death of Neto; succeeded by José Eduardo dos Santos.
1980 Constitution amended to provide for an elected people's assembly. UNITA guerrillas, aided by South Africa, continued raids against the Luanda government and bases of the South West Africa People's Organization (SWAPO) in Angola.
1984 South Africa promised to withdraw its forces if the Luanda government guaranteed that areas vacated would not be filled by Cuban or SWAPO units (the Lusaka Agreement).
1985 South African forces officially withdrawn.
1986 Further South African raids into Angola. UNITA continuing to receive South African support.
1988 Peace treaty, providing for the withdrawal of all foreign troops, signed with South Africa and Cuba.
1989 Cease-fire with UNITA broke down; guerrilla activity resumed
1990 Peace offer by rebels. Return to multiparty politics promised.
1992 MPLA–PT's general-election victory fiercely disputed by UNITA, plunging the country into renewed civil war. UNITA accepted seats in the new government, but fighting continued.
1993 Continued fighting posed the threat of return to civil war.

Antigua and Barbuda
State of

area Antigua 280 sq km/108 sq mi, Barbuda 161 sq km/62 sq mi, plus
Redonda 1 sq km/0.4 sq mi
capital and chief port St John's
towns Codrington (on Barbuda)
physical low-lying tropical islands of limestone and coral with some
higher volcanic outcrops; no rivers and low rainfall result in frequent
droughts and deforestation
head of state Elizabeth II from 1981 represented by governor general
head of government Vere C Bird from 1981
political system liberal democracy
exports sea-island cotton, rum, lobsters
currency Eastern Caribbean dollar
population (1989) 83,500; growth rate 1.3% p.a.
life expectancy 70 years
language English
religion Christian (mostly Anglican)
literacy 90% (1985)
GDP $173 million (1985); $2,200 per head
chronology

1493 Antigua visited by Christopher Columbus.

1632 Antigua colonized by English settlers.

1667 Treaty of Breda formally ceded Antigua to Britain.

1871–1956 Antigua and Barbuda administered as part of the Leeward Islands federation.

1967 Antigua and Barbuda became an associated state within the Commonwealth, with full internal independence.

1971 Progressive Labour Movement (PLM) won the general election by defeating the Antigua Labour Movement (ALP).

1976 PLM called for early independence, but ALP urged caution. ALP won the general election.

1981 Independence from Britain achieved.

1983 Assisted US invasion of Grenada.

1984 ALP won a decisive victory in the general election.

1985 ALP re-elected.

1989 Another sweeping general election victory for the ALP under Vere Bird.

1991 Bird remained in power despite calls for his resignation.

Argentina
Republic of
(*República
Argentina*)

area 2,780,092 sq km/1,073,116 sq mi
capital Buenos Aires (to move to Viedma)
towns Rosario, Córdoba, Tucumán, Mendoza, Santa Fé, La Plata
physical mountains in W, forest and savanna in N, pampas (treeless
plains) in E central area, Patagonian plateau in S; rivers Colorado,
Salado, Paraná, Uruguay, Río de la Plata estuary
territories part of Tierra del Fuego
environment an estimated 20,000 sq km/7,700 sq mi of land has been
swamped with salt water
head of state and government Carlos Menem from 1989
political system emergent democratic federal republic
exports livestock products, cereals, wool, tannin, peanuts, minerals
(coal, copper, molybdenum, gold, silver, lead, zinc, barium,
uranium); huge resources of oil, natural gas, hydroelectric power
currency peso = 10,000 australs (which it replaced 1992)
population (1990 est) 32,686,000; growth rate 1.5% p.a.

life expectancy men 66, women 73
languages Spanish (official); English, Italian, German, French
religion Roman Catholic (state-supported)
literacy men 96%, women 95% (1985 est)
GDP $70.1 bn (1990); $2,162 per head
chronology
1816 Independence achieved from Spain, followed by civil wars.
1946 Juan Perón elected president, supported by his wife 'Evita'.
1952 'Evita' Perón died.
1955 Perón overthrown and civilian administration restored.
1966 Coup brought back military rule.
1973 Perón returned from exile in Spain as president, with his third wife, Isabel, as vice president.
1974 Perón died, succeeded by Isabel.
1976 Coup resulted in rule by a military junta led by Lt-Gen Jorge Videla. Congress dissolved, and hundreds of people detained.
1976–83 Ferocious campaign ('dirty war') against left-wing elements.
1978 Videla retired. Succeeded by General Roberto Viola, who promised a return to democracy.
1981 Viola died suddenly. Replaced by General Leopoldo Galtieri.
1982 With a deteriorating economy, Galtieri sought popular support by ordering an invasion of the British-held Falkland Islands. After losing the short war, Galtieri was replaced by General Reynaldo Bignone.
1983 Amnesty law passed and democratic constitution of 1853 revived. General elections won by Raúl Alfonsín and his party. Armed forces under scrutiny.
1984 National Commission on the Disappearance of Persons (CONADEP) reported on over 8,000 people who had disappeared during the 'dirty war' of 1976–83.
1985 A deteriorating economy forced Alfonsín to seek help from the International Monetary Fund and introduce an austerity programme.
1988 Unsuccessful army coup.
1989 Carlos Menem, of the Justicialist Party, elected president.
1990 Full diplomatic relations with the UK restored. Menem elected Justicialist Party leader. Revolt by army officers thwarted.
1992 New currency introduced.

Armenia
Republic of

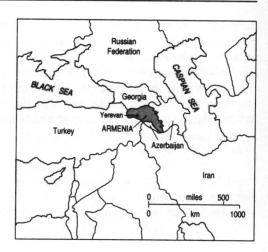

area 29,800 sq km/11,500 sq mi
capital Yerevan
towns Kumayri (formerly Leninakan)
physical mainly mountainous (including Mount Ararat), wooded
head of state Levon Ter-Petrossian from 1990
head of government Gagik Arutyunyan from 1991
political system emergent democracy
products copper, molybdenum, cereals, cotton, silk
population (1991) 3,580,000 (90% Armenian, 5% Azeri, 2% Russian,
2% Kurd)
language Armenian
religion traditionally Armenian Christian
chronology
1918 Became an independent republic.
1920 Occupied by the Red Army.
1936 Became a constituent republic of the USSR.
1988 Feb: demonstrations in Yerevan called for transfer of Nagorno-
Karabakh from Azerbaijan to Armenian control. Dec: earthquake
claimed around 25,000 lives and caused extensive damage.
1989 Jan–Nov: strife-torn Nagorno-Karabakh placed under

'temporary' direct rule from Moscow. Pro-autonomy Armenian National Movement founded. Nov: civil war erupted with Azerbaijan over Nagorno-Karabakh.

1990 March: Armenia boycotted USSR constitutional referendum. Aug: nationalists secured control of Armenian supreme soviet; former dissident Levon Ter-Petrossian indirectly elected president; independence declared. Nakhichevan republic affected by Nagorno-Karabakh dispute.

1991 March: overwhelming support for independence in referendum. Dec: Armenia joined new Commonwealth of Independent States (CIS); Nagorno-Karabakh declared its independence; Armenia granted diplomatic recognition by USA.

1992 Jan: admitted into Conference on Security and Cooperation in Europe (CSCE). March: joined United Nations (UN). Conflict over Nagorno-Karabakh worsened.

Australia
Commonwealth of

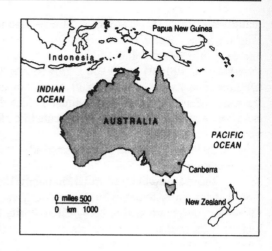

area 7,682,300 sq km/2,966,136 sq mi
capital Canberra
towns Adelaide, Brisbane, Darwin, Melbourne, Perth, Sydney
physical world's smallest, flattest, and driest continent (40% lies in the
tropics, one-third is desert, and one-third is marginal grazing); Great
Sandy, Gibson, Great Victoria, and Simpson deserts; the Great Barrier
Reef (largest coral reef in the world); Great Dividing Range and
Australian Alps in the E. The fertile SE region is watered by the
Darling, Lachlan, Murrumbridgee, and Murray rivers; rivers in the
interior are seasonal. Lake Eyre basin and Nullarbor Plain in the S
territories Norfolk Island, Christmas Island, Cocos (Keeling) Islands,
Ashmore and Cartier Islands, Coral Sea Islands, Heard Island and
McDonald Islands, Australian Antarctic Territory
environment an estimated 75% of Australia's northern tropical
rainforest has been cleared for agriculture or urban development since
Europeans first settled there in the early 19th century
head of state Elizabeth II from 1952, represented by governor general
head of government Paul Keating from 1991
political system federal constitutional monarchy
exports world's largest exporter of sheep, wool, diamonds, alumina,
coal, lead and refined zinc ores, and mineral sands; other exports

include cereals, beef, veal, mutton, lamb, sugar, nickel (world's second largest producer), iron ore
currency Australian dollar
population (1990 est) 16,650,000; growth rate 1.5% p.a.
life expectancy men 75, women 80
languages English, Aboriginal languages
religions Anglican 26%, other Protestant 17%, Roman Catholic 26%
literacy 98.5.% (1988)
GDP $220.96 bn (1988); $14,458 per head
recent chronology
1942 Achieved autonomy from UK in internal and external affairs.
1944 Liberal Party founded by Robert Menzies.
1966 Menzies resigned after being Liberal prime minister for 17 years, and was succeeded by Harold Holt.
1967 Aborigines achieved full citizenship rights.
1968 John Gorton became prime minister after Holt's death.
1971 Gorton succeeded by William McMahon, heading a Liberal–Country Party coalition.
1972 Gough Whitlam became prime minister, leading a Labor government.
1975 Senate blocked the government's financial legislation; Whitlam dismissed by the governor general, who invited Malcolm Fraser to form a Liberal–Country Party caretaker government. The action of the governor general, John Kerr, was widely criticized.
1978 Northern Territory attained self-government.
1983 Australian Labor Party returned to power under Bob Hawke.
1986 Australia Act passed by UK government, eliminating last vestiges of British legal authority in Australia.
1988 Labor foreign minister Bill Hayden appointed governor general designate. Free-trade agreement with New Zealand signed.
1990 Hawke won record fourth election victory, defeating Liberal Party by small majority.
1991 Paul Keating became new Labor Party leader and prime minister.
1992 Keating's popularity declined as economic problems continued. Oath of allegiance to British Crown abandoned.
1993 Labor Party won general election, entering fifth term of office.

Austria
Republic of
(*Republik
Österreich*)

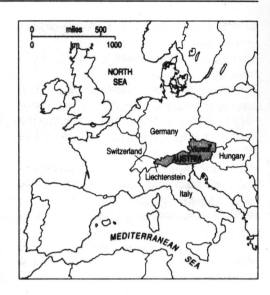

area 83,500 sq km/32,374 sq mi
capital Vienna
towns Graz, Linz, Salzburg, Innsbruck
physical landlocked mountainous state, with Alps in W and S and low
relief in E where most of the population is concentrated
environment Hainburg, the largest primeval forest left in Europe,
under threat from a dam project (suspended 1990)
head of state Thomas Klestil from 1992
head of government Franz Vranitzky from 1986
political system democratic federal republic
exports lumber, textiles, clothing, iron and steel, paper, machinery and
transport equipment, foodstuffs
currency schilling
population (1990 est) 7,595,000; growth rate 0.1% p.a.
life expectancy men 70, women 77
language German
religions Roman Catholic 85%, Protestant 6%
literacy 98% (1983)

GDP $183.3 bn (1987); $11,337 per head
chronology
1867 Emperor Franz Josef established dual monarchy of Austria–Hungary.
1914 Archduke Franz Ferdinand assassinated by a Serbian nationalist; Austria–Hungary invaded Serbia, precipitating World War I.
1918 Habsburg empire ended; republic proclaimed.
1938 Austria incorporated into German Third Reich by Hitler (the *Anschluss*).
1945 Under Allied occupation, constitution of 1920 reinstated and coalition government formed by the Socialist Party of Austria (SPÖ) and the Austrian People's Party (ÖVP).
1955 Allied occupation ended, and the independence of Austria formally recognized.
1966 ÖVP in power with Josef Klaus as chancellor.
1970 SPÖ formed a minority government, with Dr Bruno Kreisky as chancellor.
1983 Kreisky resigned and was replaced by Dr Fred Sinowatz, leading a coalition.
1986 Dr Kurt Waldheim elected president. Sinowatz resigned, succeeded by Franz Vranitzky. No party won an overall majority; Vranitzky formed a coalition of the SPÖ and the ÖVP, with ÖVP leader, Dr Alois Mock, as vice chancellor.
1989 Austria sought European Community membership.
1990 Vranitzky re-elected.
1991 Bid for EC membership endorsed by the Community.
1992 Thomas Klestil elected president, replacing Waldheim.

Azerbaijan
Republic of

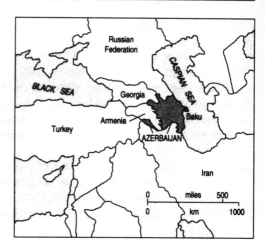

area 86,600 sq km/33,400 sq mi
capital Baku
towns Gyandzha (formerly Kirovabad), Sumgait
physical Caspian Sea; the country ranges from semidesert to the Caucasus Mountains
head of state Albulfaz Elchibey from 1992
head of government to be appointed
political system emergent democracy
products oil, iron, copper, fruit, vines, cotton, silk, carpets
population (1990) 7,145,600 (83% Azeri, 6% Russian, 6% Armenian)
language Turkic
religion traditionally Shi'ite Muslim
chronology
1917–18 A member of the anti-Bolshevik Transcaucasian Federation.
1918 Became an independent republic.
1920 Occupied by the Red Army.
1922–36 Formed part of the Transcaucasian Federal Republic with Georgia and Armenia.
1936 Became a constituent republic of the USSR.
1988 Riots followed Nagorno-Karabakh's request for transfer to Armenia.

1989 Jan–Nov: strife-torn Nagorno-Karabakh placed under
'temporary' direct rule from Moscow. Azerbaijan Popular Front
established. Nov: civil war erupted with Armenia.
1990 Jan: Soviet troops dispatched to Baku to restore order. Aug:
communists won parliamentary elections. Nakhichevan republic
affected by Nagorno-Karabakh dispute.
1991 Aug: Azeri leadership supported attempted anti-Gorbachev coup
in Moscow; independence declared. Sept: former communist Ayaz
Mutalibov elected president. Dec: joined new Commonwealth of
Independent States (CIS); Nagorno-Karabakh declared independence.
1992 Jan: admitted into Conference on Security and Cooperation in
Europe (CSCE); March: Mutalibov resigned; Azerbaijan became a
member of the United Nations (UN); accorded diplomatic recognition
by the USA. June: Albulfaz Elchibey, leader of the Popular Front,
elected president; renewed campaign against Armenia in the fight for
Nagorno-Karabakh.
1993 Prime Minister Rakham Guseinov resigned over differences with
President Elchibey.

Bahamas
Commonwealth
of the

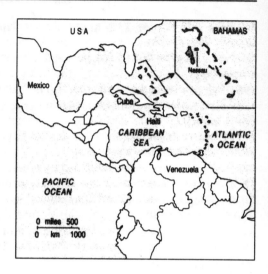

area 13,864 sq km/5,352 sq mi
capital Nassau on New Providence
towns Alice Town, Andros Town, Hope Town, Spanish Wells,
Freeport, Moss Town
physical comprises 700 tropical coral islands and about 1,000 cays
principal islands Andros, Grand Bahama, Great Abaco, Eleuthera,
New Providence, Berry Islands, Biminis, Great Inagua, Acklins,
Exumas, Mayaguana, Crooked Island, Long Island, Cat Island, Rum
Cay, Watling (San Salvador) Island
head of state Elizabeth II from 1973, represented by governor general
head of government Hubert Ingraham from 1992
political system constitutional monarchy
exports cement, pharmaceuticals, petroleum products, crawfish, salt,
aragonite, rum, pulpwood; over half the islands' employment comes
from tourism
currency Bahamian dollar
population (1990 est) 251,000; growth rate 1.8% p.a.
life expectancy men 67, women 74
languages English and some Creole

religions 29% Baptist, 23% Anglican, 22% Roman Catholic
literacy 95% (1986)
GDP $2.7 bn (1987); $11,261 per head
chronology
1964 Independence achieved from Britain.
1967 First national assembly elections; Lynden Pindling became prime minister.
1972 Constitutional conference to discuss full independence.
1973 Full independence achieved.
1983 Allegations of drug trafficking by government ministers.
1984 Deputy prime minister and two cabinet ministers resigned. Pindling denied any personal involvement and was endorsed as party leader.
1987 Pindling re-elected despite claims of frauds.
1992 Free National Movement (FNM) led by Hubert Ingraham won absolute majority in assembly elections.

Bahrain
State of
(*Dawlat al Bahrayn*)

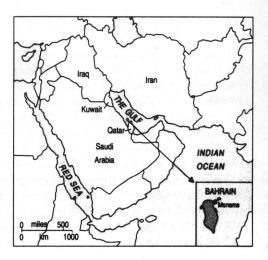

area 688 sq km/266 sq mi

capital Manama on the largest island (also called Bahrain)

towns Muharraq, Jidd Hafs, Isa Town, Mina Sulman

physical 35 islands, composed largely of sand-covered limestone; generally poor and infertile soil; flat and hot

environment a wildlife park on Bahrain preserves the endangered oryx; most of the south of the island is preserved for the ruling family's falconry

head of state and government Sheik Isa bin Sulman al-Khalifa from 1961

political system absolute emirate

exports oil, natural gas, aluminium, fish

currency Bahrain dinar

population (1990 est) 512,000 (two-thirds are nationals); growth rate 4.4% p.a.

life expectancy men 67, women 71

languages Arabic (official); Farsi, English, Urdu

religion 85% Muslim (Shi'ite 60%, Sunni 40%)

literacy men 79%, women 64% (1985 est)

GDP $3.5 bn (1987); $7,772 per head
chronology
1861 Became British protectorate.
1968 Britain announced its intention to withdraw its forces. Bahrain formed, with Qatar and the Trucial States, the Federation of Arab Emirates.
1971 Qatar and the Trucial States withdrew from the federation and Bahrain became an independent state.
1973 New constitution adopted, with an elected national assembly.
1975 Prime minister resigned and national assembly dissolved. Emir and his family assumed virtually absolute power.
1986 Gulf University established in Bahrain. A causeway was opened linking the island with Saudi Arabia.
1988 Bahrain recognized Afghan rebel government.
1991 Bahrain joined United Nations coalition that ousted Iraq from its occupation of Kuwait.

Bangladesh
People's Republic of
(*Gana Prajatantri
Bangladesh*)
(formerly *East Pakistan*)

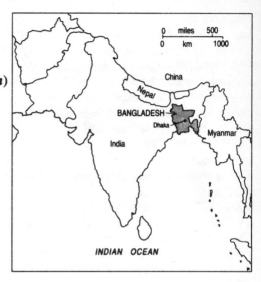

area 144,000 sq km/55,585 sq mi
capital Dhaka (formerly Dacca)
towns Chittagong, Khulna
physical flat delta of rivers Ganges (Padma) and Brahmaputra
(Jamuna), the largest estuarine delta in the world; annual rainfall of
2,540 mm/100 in; some 75% of the land is less than 3 m/10 ft above
sea level and vulnerable to flooding and cyclones; hilly in extreme
SE and NE
environment deforestation on the slopes of the Himalayas increases
the threat of flooding in the coastal lowlands, which are also subject to
devastating monsoon storms. The building of India's Farakka Barrage
has reduced the flow of the Ganges in Bangladesh and permitted salt
water to intrude further inland. Increased salinity has destroyed
fisheries, contaminated drinking water, and damaged forests
head of state Abdur Rahman Biswas from 1991
head of government Begum Khaleda Zia from 1991
political system emergent democratic republic
exports jute, tea, garments, fish products

currency taka

population (1991 est) 107,992,100; growth rate 2.17% p.a.; just over 1 million people live in small ethnic groups in the tropical Chittagong Hill Tracts, Mymensingh, and Sylhet districts

life expectancy men 50, women 52

language Bangla (Bengali)

religions Sunni Muslim 85%, Hindu 14%

literacy men 43%, women 22% (1985 est)

GDP $17.6 bn (1987); $172 per head

chronology

1947 Formed into eastern province of Pakistan on partition of British India.

1970 Half a million killed in flood.

1971 Bangladesh emerged as independent nation, under leadership of Sheik Mujibur Rahman, after civil war.

1975 Mujibur Rahman assassinated. Martial law imposed.

1976–77 Maj-Gen Zia ur-Rahman assumed power.

1978–79 Elections held and civilian rule restored.

1981 Assassination of Maj-Gen Zia.

1982 Lt-Gen Ershad assumed power in army coup. Martial law reimposed.

1986 Elections held but disputed. Martial law ended.

1987 State of emergency declared in response to opposition demonstrations.

1988 Assembly elections boycotted by main opposition parties. State of emergency lifted. Islam made state religion. Monsoon floods left 30 million homeless and thousands dead.

1989 Power devolved to Chittagong Hill Tracts to end 14-year conflict between local people and army-protected settlers.

1990 Following mass antigovernment protests, President Ershad resigned; Shahabuddin Ahmad became interim president.

1991 Feb: elections resulted in coalition government with Bangladesh Nationalist Party (BNP) dominant. April: cyclone killed around 139,000 and left up to 10 million homeless. Sept: parliamentary government restored; Abdur Rahman Biswas elected president.

Barbados

area 430 sq km/166 sq mi
capital Bridgetown
towns Speightstown, Holetown, Oistins
physical most easterly island of the West Indies; surrounded by coral
reefs; subject to hurricanes June–Nov
head of state Elizabeth II from 1966, represented by governor general
Hugh Springer from 1984
head of government prime minister Erskine Lloyd Sandiford from
1987
political system constitutional monarchy
exports sugar, rum, electronic components, clothing, cement
currency Barbados dollar
population (1990 est) 260,000; growth rate 0.5% p.a.
life expectancy men 70, women 75
languages English and Bajan (Barbadian English dialect)
religions 70% Anglican, 9% Methodist, 4% Roman Catholic
literacy 99% (1984)
GDP $1.4 bn (1987); $5,449 per head

chronology

1627 Became British colony; developed as a sugar-plantation economy, initially on basis of slavery.

1834 Slaves freed.

1951 Universal adult suffrage introduced. Barbados Labour Party (BLP) won general election.

1954 Ministerial government established.

1961 Independence achieved from Britain. Democratic Labour Party (DLP), led by Errol Barrow, in power.

1966 Barbados achieved full independence within Commonwealth. Barrow became the new nation's first prime minister.

1972 Diplomatic relations with Cuba established.

1976 BLP, led by Tom Adams, returned to power.

1983 Barbados supported US invasion of Grenada.

1985 Adams died; Bernard St John became prime minister.

1986 DLP, led by Barrow, returned to power.

1987 Barrow died; Erskine Lloyd Sandiford became prime minister.

1989 New National Democratic Party (NDP) opposition formed.

1991 DLP, under Erskine Sandiford, won general election.

Belarus
Republic of

area 207,600 sq km/80,100 sq mi
capital Minsk (Mensk)
towns Gomel, Vitebsk, Mogilev, Bobruisk, Grodno, Brest
physical more than 25% forested; rivers W Dvina, Dnieper and its
tributaries, including the Pripet and Beresina; the Pripet Marshes in the
E; mild and damp climate
environment large areas contaminated by fallout from Chernobyl
head of state Stanislav Shushkevich from 1991
head of government Vyacheslav Kebich from 1990
political system emergent democracy
products peat, agricultural machinery, fertilizers, glass, textiles,
leather, salt, electrical goods, meat, dairy produce
currency rouble and dukat
population (1990) 10,200,000 (77% Byelorussian 'Eastern Slavs',
13% Russian, 4% Polish, 1% Jewish)
languages Byelorussian, Russian
religions Roman Catholic, Russian Orthodox, with Baptist and
Muslim minorities

chronology

1918–19 Briefly independent from Russia.

1937–41 More than 100,000 people were shot in mass executions ordered by Stalin.

1941–44 Occupied by Nazi Germany.

1945 Became a founding member of the United Nations.

1986 April: fallout from the Chernobyl nuclear reactor in Ukraine contaminated a large area.

1989 Byelorussian Popular Front established as well as a more extreme nationalist organization, the Tolaka group.

1990 Sept: Byelorussian established as state language and republican sovereignty declared.

1991 April: Minsk hit by nationalist-backed general strike, calling for disbandment of Communist Party (CP) workplace cells. Aug: declared independence from Soviet Union in wake of failed anti-Gorbachev coup; CP suspended. Sept: Shushkevich elected president. Dec: Commonwealth of Independent States (CIS) formed in Minsk; Belarus accorded diplomatic recognition by USA.

1992 Jan: admitted into Conference on Security and Cooperation in Europe (CSCE). May: protocols signed with USA agreeing to honour START disarmament treaty.

Belgium
Kingdom of
(French *Royaume de
Belgique*, Flemish
Koninkrijk België)

area 30,510 sq km/11,784 sq mi
capital Brussels
towns Ghent, Liège, Charleroi, Bruges, Antwerp, Ostend
physical fertile coastal plain in NW, central rolling hills rise eastwards,
hills and forest in SE
environment a 1989 government report judged the drinking water in
Flanders to be 'seriously substandard'
head of state King Baudouin from 1951
head of government Jean-Luc Dehaene from 1992
political system liberal democracy
exports iron, steel, textiles, manufactured goods, petrochemicals,
plastics, vehicles, diamonds
currency Belgian franc
population (1990 est) 9,895,000 (comprising Flemings and Walloons);
growth rate 0.1% p.a.
life expectancy men 72, women 78
languages in the N (Flanders) Flemish 55%; in the S (Wallonia)
Walloon (a French dialect) 32%; bilingual 11%; German (E border)

0.6%; all are official
religion Roman Catholic 75%
literacy 98% (1984)
GDP $111 bn (1986); $9,230 per head
chronology
1830 Belgium became an independent kingdom.
1914 Invaded by Germany.
1940 Again invaded by Germany.
1948 Belgium became founding member of Benelux Customs Union.
1949 Belgium became founding member of Council of Europe and
NATO.
1951 Leopold III abdicated in favour of his son Baudouin.
1957 Belgium became founding member of the European Economic
Community.
1971 Steps towards regional autonomy taken.
1972 German-speaking members included in the cabinet for the
first time.
1973 Linguistic parity achieved in government appointments.
1974 Leo Tindemans became prime minister. Separate regional
councils and ministerial committees established.
1978 Wilfried Martens succeeded Tindemans as prime minister.
1980 Open violence over language divisions. Regional assemblies for
Flanders and Wallonia and a three-member executive for Brussels
created.
1981 Short-lived coalition led by Mark Eyskens was followed by the
return of Martens.
1987 Martens head of caretaker government after break-up of
coalition.
1988 Following a general election, Martens formed a new coalition
between the Flemish Christian Social Party (CVP), French Socialist
Party (PS), Flemish Socialist Party (SP), French Social Christian Party
(PSC), and the Flemish People's Party (VU).
1992 Martens-led coalition collapsed; Jean-Luc Dehaene formed a
new CVP-led coalition. It was announced that a federal system would
be introduced.

Belize
(formerly *British Honduras*)

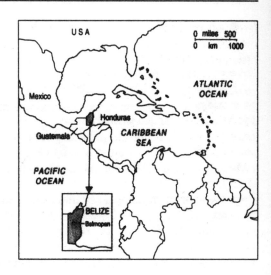

area 22,963 sq km/8,864 sq mi
capital Belmopan
towns Belize City, Dangriga, Punta Gorda, Orange Walk, Corozal
physical tropical swampy coastal plain, Maya Mountains in S; over 90% forested
environment since 1981 Belize has developed an extensive system of national parks and reserves to protect large areas of tropical forest, coastal mangrove, and offshore islands. Forestry has been replaced by agriculture and ecotourism, which are now the most important sectors of the economy; world's first jaguar reserve created 1986 in the Cockscomb Mountains
head of state Elizabeth II from 1981, represented by governor general
head of government George Price from 1989
political system constitutional monarchy
exports sugar, citrus fruits, rice, fish products, bananas
currency Belize dollar
population (1990 est) 180,400 (including Mayan minority in the interior); growth rate 2.5% p.a.
life expectancy (1988) 60 years

languages English (official); Spanish (widely spoken), native Creole dialects
religions Roman Catholic 60%, Protestant 35%
literacy 93% (1988)
GDP $247 million (1988); $1,220 per head
chronology
1862 Belize became a British colony.
1954 Constitution adopted, providing for limited internal self-government. General election won by George Price.
1964 Self-government achieved from the UK (universal adult suffrage introduced).
1965 Two-chamber national assembly introduced, with Price as prime minister.
1970 Capital moved from Belize City to Belmopan.
1973 British Honduras became Belize.
1975 British troops sent to defend the disputed frontier with Guatemala.
1977 Negotiations undertaken with Guatemala but no agreement reached.
1980 United Nations called for full independence.
1981 Full independence achieved. Price became prime minister.
1984 Price defeated in general election. Manuel Esquivel formed the government. The UK reaffirmed its undertaking to defend the frontier.
1989 Price and the People's United Party (PUP) won the general election.
1991 Diplomatic relations with Guatemala established.

Benin
People's Republic of
(*République*
Populaire du Bénin)

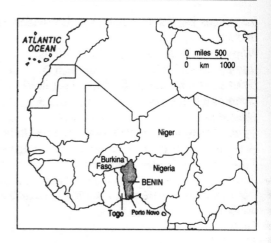

area 112,622 sq km/43,472 sq mi
capital Porto Novo (official), Cotonou (de facto)
towns Abomey, Natitingou, Parakou, Cotonou
physical flat to undulating terrain; hot and humid in S; semi-arid in N
head of state and government Nicéphore Soglo from 1991
political system socialist pluralist republic
exports cocoa, peanuts, cotton, palm oil, petroleum, cement
currency CFA franc
population (1990 est) 4,840,000; growth rate 3% p.a.
life expectancy men 42, women 46
languages French (official); Fon 47% and Yoruba 9% in south; six
major tribal languages in north
religions animist 65%, Christian 17%, Muslim 13%
literacy men 37%, women 16% (1985 est)
GDP $1.6 bn (1987); $365 per head
chronology
1851 Under French control.
1958 Became self-governing dominion within the French Community.
1960 Independence achieved from France.
1960–72 Acute political instability, with switches from civilian to
military rule.

1972 Military regime established by General Mathieu Kerekou.
1974 Kerekou announced that the country would follow a path of 'scientific socialism'.
1975 Name of country changed from Dahomey to Benin.
1977 Return to civilian rule under a new constitution.
1980 Kerekou formally elected president by the national revolutionary assembly.
1989 Marxist-Leninism dropped as official ideology. Strikes and protests against Kerekou's rule mounted; demonstrations banned and army deployed against protesters.
1990 Referendum support for multiparty politics.
1991 Multiparty elections held. Kerekou defeated in presidential elections by Nicéphore Soglo.

Bhutan
Kingdom of
(*Druk-yul*)

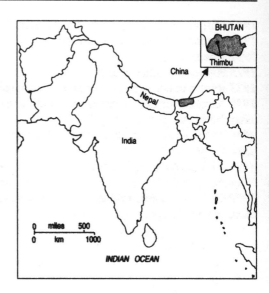

area 46,500 sq km/17,954 sq mi
capital Thimbu (Thimphu)
towns Paro, Punakha, Mongar
physical occupies southern slopes of the Himalayas; cut by valleys
formed by tributaries of the Brahmaputra; thick forests in S
head of state and government Jigme Singye Wangchuk from 1972
political system absolute monarchy
exports timber, talc, fruit and vegetables, cement, distilled spirits,
calcium carbide
currency ngultrum; also Indian currency
population (1990 est) 1,566,000; growth rate 2% p.a. (75% Ngalops
and Sharchops, 25% Nepalese)
life expectancy men 44, women 43
languages Dzongkha (official, a Tibetan dialect), Sharchop, Bumthap,
Nepali, and English
religions 75% Lamaistic Buddhist (state religion), 25% Hindu
literacy 5%
GDP $250 million (1987); $170 per head

chronology

1865 Trade treaty with Britain signed.

1907 First hereditary monarch installed.

1910 Anglo-Bhutanese Treaty signed.

1949 Indo-Bhutan Treaty of Friendship signed.

1952 King Jigme Dorji Wangchuk installed.

1953 National assembly established.

1959 4,000 Tibetan refugees given asylum.

1968 King established first cabinet.

1972 King died and was succeeded by his son Jigme Singye Wangchuk.

1979 Tibetan refugees told to take Bhutanese citizenship or leave; most stayed.

1983 Bhutan became a founding member of the South Asian Regional Association for Cooperation (SAARC).

1988 King imposed 'code of conduct' suppressing Nepalese customs.

1990 Hundreds of people allegedly killed during prodemocracy demonstrations.

Bolivia
Republic of
(*República de
Bolivia*)

area 1,098,581 sq km/424,052 sq mi
capital La Paz (seat of government), Sucre (legal capital)
towns Santa Cruz, Cochabamba, Oruro, Potosí
physical high plateau (Altiplano) between mountain ridges
(cordilleras); forest and lowlands (llano) in the E
head of state and government Jaime Paz Zamora from 1989
political system emergent democratic republic
exports tin, antimony (second largest world producer), other
nonferrous metals, oil, gas, coffee, sugar, cotton
currency boliviano
population (1990 est) 6,730,000; (Quechua 25%, Aymara 17%,
mestizo (mixed) 30%, European 14%); growth rate 2.7% p.a.
life expectancy men 51, women 54
languages Spanish, Aymara, Quechua (all official)
religion Roman Catholic 95% (state-recognized)
literacy men 84%, women 65% (1985 est)

GDP $4.2 bn (1987); $617 per head
chronology
1825 Liberated from Spanish rule by Simón Bolívar; independence achieved (formerly known as Upper Peru).
1952 Dr Víctor Paz Estenssoro elected president.
1956 Dr Hernán Siles Zuazo became president.
1960 Estenssoro returned to power.
1964 Army coup led by the vice president, General René Barrientos.
1966 Barrientos became president.
1967 Uprising, led by 'Che' Guevara, put down with US help.
1969 Barrientos killed in plane crash, replaced by Vice President Siles Salinas. Army coup deposed him.
1970 Army coup put General Juan Torres González in power.
1971 Torres replaced by Col Hugo Banzer Suárez.
1973 Banzer promised a return to democratic government.
1974 Attempted coup prompted Banzer to postpone elections and ban political and trade-union activity.
1978 Elections declared invalid after allegations of fraud.
1980 More inconclusive elections followed by another coup, led by General Luis García. Allegations of corruption and drug trafficking led to cancellation of US and EC aid.
1981 García replaced by General Celso Torrelio Villa.
1982 Torrelio resigned. Replaced by military junta led by General Guido Vildoso. Because of worsening economy, Vildoso asked congress to install a civilian administration. Dr Siles Zuazo chosen as president.
1983 Economic aid from USA and Europe resumed.
1984 New coalition government formed by Siles. Abduction of president by right-wing officers. The president undertook a five-day hunger strike as an example to the nation.
1985 President Siles resigned. Election result inconclusive; Dr Paz Estenssoro, at the age of 77, chosen by congress as president.
1989 Jaime Paz Zamora, Movement of the Revolutionary Left (MIR) elected president in power-sharing arrangement with Hugo Banzer Suárez, pledged to preserve free-market policies.
1992 The new Solidarity Civil Union (UCS) party gained support.

Bosnia-Herzegovina
Republic of

area 51,129 sq km/19,745 sq mi
capital Sarajevo
towns Banja Luka, Mostar, Prijedor, Tuzla, Zenica
physical barren, mountainous country
population (1990) 4,300,000 including 44% Muslims, 33% Serbs,
17% Croats; a complex patchwork of ethnically mixed communities
head of state Alija Izetbegović from 1990
head of government Mile Akmadzic from 1992
political system emergent democracy
products citrus fruits and vegetables; iron, steel, and leather goods;
textiles
language Serbian variant of Serbo-Croatian
religions Sunni Muslim, Serbian Orthodox, Roman Catholic
chronology
1918 Incorporated in the future Yugoslavia.
1941 Occupied by Nazi Germany.
1945 Became republic within Yugoslav Socialist Federation.
1980 Upsurge in Islamic nationalism.

1990 Ethnic violence erupted between Muslims and Serbs. Nov–Dec: communists defeated in multiparty elections; coalition formed by Serb, Muslim, and Croatian parties.

1991 May: Serbia–Croatia conflict spread disorder into Bosnia-Herzegovina. Aug: Serbia revealed plans to annex the SE part of the republic. Sept: Serbian enclaves established by force. Oct: 'sovereignty' declared. Nov: plebiscite by Serbs favoured remaining within Yugoslavia; Serbs and Croats established autonomous communities.

1992 Feb–March: Muslims and Croats voted overwhelmingly in favour of independence; referendum boycotted by Serbs. April: USA and EC recognized Bosnian independence. Ethnic hostilities escalated, with Serb forces occupying E and Croatian forces much of W; state of emergency declared; all-out civil war ensued. May: admitted to United Nations. June: Canadian–French UN forces drafted into Sarajevo to break three-month siege of city by Serbs. July: Canadian forces replaced by French, Egyptians, and Ukrainians. Official cease-fire broken intermittently by both sides; UN and EC mediators vainly sought truce. Fighting continued, with accusations of 'ethnic cleansing' being practised, particularly by Serbs. Oct: UN Security Council voted to create a war crimes commission and imposed ban on military flights over Bosnia-Herzegovina. First British troops deployed.

1993 Jan: UN–EC peace plan, proposing to divide country into 10 autonomous, ethnically-controlled provinces, accepted in principle by Serbs and Croats but fighting continued. March: USA began airdrops of food and medical supplies.

Botswana
Republic of

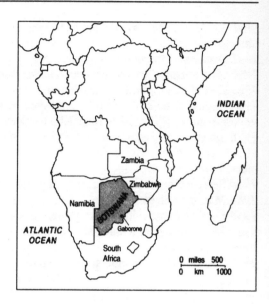

area 582,000 sq km/225,000 sq mi
capital Gaborone
towns Mahalpye, Serowe, Tutume, Francistown
physical desert in SW, plains in E, fertile lands and swamp in N
environment the Okavango Swamp is threatened by plans to develop
the area for mining and agriculture
head of state and government Quett Ketamile Joni Masire from 1980
political system democratic republic
exports diamonds (third largest producer in world), copper, nickel,
meat products, textiles
currency pula
population (1990 est) 1,218,000 (Bamangwato 80%, Bangwaketse
20%); growth rate 3.5% p.a.
life expectancy (1988) 59 years
languages English (official), Setswana (national)
religions Christian 50%, animist 50%
literacy (1988) 84%

GDP $2.0 bn (1988); $1,611 per head
chronology
1885 Became a British protectorate.
1960 New constitution created a legislative council.
1963 End of rule by High Commission.
1965 Capital transferred from Mafeking to Gaborone. Internal self-government achieved. Sir Seretse Khama elected head of government.
1966 Independence achieved from Britain. New constitution came into effect; name changed from Bechuanaland to Botswana; Seretse Khama elected president.
1980 Seretse Khama died; succeeded by Vice President Quett Masire.
1984 Masire re-elected.
1985 South African raid on Gaborone.
1987 Joint permanent commission with Mozambique established, to improve relations.
1989 The Botswana Democratic Party (BDP) and Masire re-elected.

Brazil
Federative
Republic of
(*República
Federativa do
Brasil*)

area 8,511,965 sq km/3,285,618 sq mi
capital Brasília
towns São Paulo, Belo Horizonte, Curitiba, Fortaleza, Rio de Janeiro,
Recife, Salvador
physical the densely forested Amazon basin covers the northern half
of the country with a network of rivers; the south is fertile
environment Brazil has one-third of the world's tropical rainforest. It
contains 55,000 species of flowering plants (the greatest variety in the
world) and 20% of all the world's bird species. During the 1980s at
least 7% of the Amazon rainforest was destroyed by settlers who
cleared the land for cultivation and grazing
head of state and government Itamar Franco from 1992
political system emergent democratic federal republic
exports coffee, sugar, soya beans, cotton, textiles, timber, motor
vehicles, iron, chrome, manganese, tungsten and other ores; the
world's sixth largest arms exporter

currency cruzado (introduced 1986); inflation 1990 was 1,795%
population (1990 est) 153,770,000 (including 200,000 Indians, mostly living on reservations); growth rate 2.2% p.a.
life expectancy men 61, women 66
languages Portuguese (official); 120 Indian languages
religions Roman Catholic 89%; Indian faiths
literacy men 79%, women 76% (1985 est)
GDP $352 bn (1988); $2,434 per head
chronology
1822 Independence achieved from Portugal.
1889 Monarchy abolished and republic established.
1891 Constitution for a federal state adopted.
1960 Capital moved to Brasília from Rio de Janeiro.
1961 João Goulart became president.
1964 Bloodless coup made General Castelo Branco president; he assumed dictatorial powers, abolishing free political parties.
1967 New constitution adopted. Branco succeeded by Marshal da Costa e Silva.
1969 Da Costa e Silva resigned and a military junta took over.
1974 General Ernesto Geisel became president.
1978 General Baptista de Figueiredo became president.
1979 Political parties legalized again.
1984 Mass calls for a return to fully democratic government.
1985 Tancredo Neves became first civilian president in 21 years. Neves died and was succeeded by the vice president, José Sarney.
1988 New constitution approved, transferring power from the president to the congress. Measures announced to halt large-scale burning of Amazonian rainforest for cattle grazing.
1989 Forest Protection Service and Ministry for Land Reform abolished. International concern over how much of the Amazon has been burned. Fernando Collor, National Reconstruction Party (PRN), elected president, pledging free-market economic policies.
1990 Government won the general election offset by mass abstentions.
1992 June: Earth Summit, global conference on the environment, held in Rio de Janeiro. Sept: Collor charged with corruption and stripped of his powers. Replaced by Vice President Itamar Franco.

Brunei
Islamic Sultanate of
(*Negara Brunei
Darussalam*)

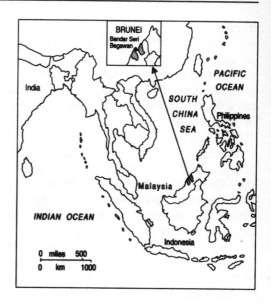

area 5,765 sq km/2,225 sq mi
capital Bandar Seri Begawan
towns Tutong, Seria, Kuala Belait
physical flat coastal plain with hilly lowland in W and mountains in E;
75% of the area is forested; the Limbang valley splits Brunei in two,
and its cession to Sarawak 1890 is disputed by Brunei
head of state and of government HM Muda Hassanal Bolkiah
Mu'izzaddin Waddaulah, Sultan of Brunei, from 1968
political system absolute monarchy
exports liquefied natural gas (world's largest producer) and oil, both
expected to be exhausted by the year 2000
currency Brunei dollar
population (1990 est) 372,000 (65% Malay, 20% Chinese – few
Chinese granted citizenship); growth rate 12% p.a.
life expectancy 74 years
languages Malay (official), Chinese (Hokkien), English
religion 60% Muslim (official)

literacy 95%
GDP $3.4 bn (1985); $20,000 per head
chronology
1888 Brunei became a British protectorate.
1941–45 Occupied by Japan.
1959 Written constitution made Britain responsible for defence and external affairs.
1962 Sultan began rule by decree.
1963 Proposal to join Malaysia abandoned.
1967 Sultan abdicated in favour of his son, Hassanal Bolkiah.
1971 Brunei given internal self-government.
1975 United Nations resolution called for independence for Brunei.
1984 Independence achieved from Britain, with Britain maintaining a small force to protect the oil- and gasfields.
1985 A 'loyal and reliable' political party, the Brunei National Democratic Party (BNDP), legalized.
1986 Death of former sultan, Sir Omar. Formation of multiethnic Brunei National United Party (BNUP).
1988 BNDP banned.

Bulgaria
Republic of
(*Republika
Bulgaria*)

area 110,912 sq km/42,812 sq mi
capital Sofia
towns Plovdiv, Ruse, Burgas, Varna
physical lowland plains in N and SE separated by mountains that
cover three-quarters of the country
environment pollution has virtually eliminated all species of fish once
caught in the Black Sea. Vehicle-exhaust emissions in Sofia have led to
dust concentrations more than twice the medically accepted level
head of state Zhelyu Zhelev from 1990
head of government Lyuben Berov from 1992
political system emergent democratic republic
exports textiles, leather, chemicals, nonferrous metals, timber,
machinery, tobacco, cigarettes (world's largest exporter)
currency lev
population (1990 est) 8,978,000 (including 900,000–1,500,000 ethnic
Turks, concentrated in S and NE); growth rate 0.1% p.a.
life expectancy men 69, women 74
languages Bulgarian, Turkish

religions Eastern Orthodox Christian 90%, Sunni Muslim 10%
literacy 98%
GDP $25.4 bn (1987); $2,836 per head
chronology
1908 Bulgaria became a kingdom independent of Turkish rule.
1944 Soviet invasion of German-occupied Bulgaria.
1946 Monarchy abolished and communist-dominated people's
republic proclaimed.
1947 Soviet-style constitution adopted.
1949 Death of Georgi Dimitrov, the communist government leader.
1954 Election of Todor Zhivkov as Communist Party general
secretary; made nation a loyal satellite of USSR.
1971 Constitution modified; Zhivkov elected president.
1985–89 Large administrative and personnel changes made
haphazardly under Soviet stimulus.
1987 New electoral law introduced multicandidate elections.
1989 Programme of 'Bulgarianization' resulted in mass exodus of
Turks to Turkey. Nov: Zhivkov ousted by Petar Mladenov.
Dec: opposition parties allowed to form.
1990 April: Bulgarian Communist Party (BCP) renamed Bulgarian
Socialist Party (BSP). Aug: Dr Zhelyu Zhelev elected president. Nov:
government headed by Andrei Lukanov resigned, replaced Dec by
coalition led by Dimitur Popov.
1991 July: new constitution adopted. Oct: Union of Democratic Forces
(UDF) beat BSP in general election by narrow margin; formation of
first noncommunist, UDF-minority government under Filip Dimitrov.
1992 Zhelev became Bulgaria's first directly elected president.
Relations with West greatly improved. Dimitrov resigned after vote of
no confidence; replaced by Lyuben Berov.

Burkina Faso

The People's
Democratic
Republic of
(formerly
Upper Volta)

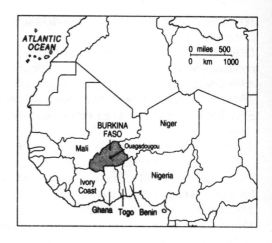

area 274,122 sq km/105,811 sq mi
capital Ouagadougou
towns Bobo-Dioulasso, Koudougou
physical landlocked plateau with hills in W and SE; headwaters of the
river Volta; semi-arid in N, forest and farmland in S
environment tropical savanna subject to overgrazing and deforestation
head of state and government Blaise Compaoré from 1987
political system transitional
exports cotton, karite nuts, groundnuts, livestock, hides, skins, sesame,
cereals
currency CFA franc
population (1990 est) 8,941,000; growth rate 2.4% p.a.
life expectancy men 44, women 47
languages French (official); about 50 native Sudanic languages
spoken by 90% of population
religions animist 53%, Sunni Muslim 36%, Roman Catholic 11%
literacy men 21%, women 6% (1985 est)
GDP $1.6 bn (1987); $188 per head
chronology
1958 Became a self-governing republic within the French Community.

1960 Independence from France, with Maurice Yaméogo as the first president.

1966 Military coup led by Col Lamizana. Constitution suspended, political activities banned, and a supreme council of the armed forces established.

1969 Ban on political activities lifted.

1970 Referendum approved a new constitution leading to a return to civilian rule.

1974 After experimenting with a mixture of military and civilian rule, Lamizana reassumed full power.

1977 Ban on political activities removed. Referendum approved a new constitution based on civilian rule.

1978 Lamizana elected president.

1980 Lamizana overthrown in bloodless coup led by Col Zerbo.

1982 Zerbo ousted in a coup by junior officers. Major Ouédraogo became president and Thomas Sankara prime minister.

1983 Sankara seized complete power.

1984 Upper Volta renamed Burkina Faso, 'land of upright men'.

1987 Sankara killed in coup led by Blaise Compaoré.

1989 New government party Organization for Popular Democracy–Workers' Party (ODP–MT) formed by merger of other pro-government parties. Coup against Compaoré foiled.

1991 New constitution approved. Compaoré re-elected president.

1992 Multiparty elections won by FP–Popular Front.

Burundi
Republic of
(*Republika
y'Uburundi*)

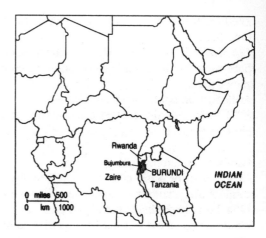

area 27,834 sq km/10,744 sq mi
capital Bujumbura
towns Gitega, Bururi, Ngozi, Muyinga
physical landlocked grassy highland straddling watershed of Nile
and Congo
head of state and government Pierre Buyoya from 1987
political system one-party military republic
exports coffee, cotton, tea, nickel, hides, livestock, cigarettes, beer,
soft drinks; there are 500 million tonnes of peat reserves in the basin of
the Akanyaru River
currency Burundi franc
population (1990 est) 5,647,000 (of whom 15% are the Nilotic Tutsi,
still holding most of the land and political power, 1% are Pygmy Twa,
and the remainder Bantu Hutu); growth rate 2.8% p.a.
life expectancy men 45, women 48
languages Kirundi (a Bantu language) and French (both official),
Kiswahili
religions Roman Catholic 62%, Protestant 5%, Muslim 1%,
animist 32%
literacy men 43%, women 26% (1985)
GDP $1.1 bn (1987); $230 per head

chronology

1962 Separated from Ruanda-Urundi, as Burundi, and given independence as a monarchy under King Mwambutsa IV.

1966 King deposed by his son Charles, who became Ntare V; he was in turn deposed by his prime minister, Capt Michel Micombero, who declared Burundi a republic.

1972 Ntare V killed, allegedly by the Hutu ethnic group. Massacres of 150,000 Hutus by the rival Tutsi ethnic group, of which Micombero was a member.

1973 Micombero made president and prime minister.

1974 Union for National Progress (UPRON) declared the only legal political party, with the president as its secretary general.

1976 Army coup deposed Micombero. Col Jean-Baptiste Bagaza appointed president by the Supreme Revolutionary Council.

1981 New constitution adopted, providing for a national assembly.

1984 Bagaza elected president as sole candidate.

1987 Bagaza deposed in coup Sept. Maj Pierre Buyoya headed new Military Council for National Redemption.

1988 Some 24,000 majority Hutus killed by Tutsis.

1992 New constitution approved.

Cambodia
State of (formerly
Khmer Republic
1970–76,
*Democratic
Kampuchea*
1976–79,
*People's Republic of
Kampuchea*
1979–89)

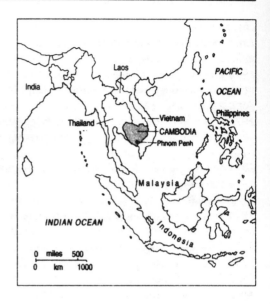

area 181,035 sq km/69,880 sq mi
capital Phnom Penh
towns Battambang, Kompong Som
physical mostly flat forested plains with mountains in SW and N;
Mekong River runs N–S
head of state Prince Norodom Sihanouk from 1991
head of government Hun Sen from 1985
political system transitional
exports rubber, rice, pepper, wood, cattle
currency Cambodian riel
population (1990 est) 6,993,000; growth rate 2.2% p.a.
life expectancy men 42, women 45
languages Khmer (official), French
religion Theravāda Buddhist 95%
literacy men 78%, women 39% (1980 est)
GDP $592 mn (1987); $83 per head

chronology

1863–1941 French protectorate.

1941–45 Occupied by Japan.

1946 Recaptured by France.

1953 Independence achieved from France.

1970 Prince Sihanouk overthrown by US-backed Lon Nol.

1975 Lon Nol overthrown by Khmer Rouge.

1976–78 Khmer Rouge introduced an extreme communist programme, forcing urban groups into rural areas and bringing about over 2.5 million deaths from famine, disease, and maltreatment.

1978–79 Vietnamese invasion and installation of Heng Samrin government.

1982 The three main anti-Vietnamese resistance groups formed an alliance under Prince Sihanouk.

1987 Vietnamese troop withdrawal began.

1989 Sept: completion of Vietnamese withdrawal. Nov: United Nations peace proposal rejected by Phnom Penh government.

1991 Oct: Peace agreement signed in Paris, providing for a UN Transitional Authority in Cambodia (UNTAC) to administer country in transition period in conjunction with all-party Supreme National Council; communism abandoned. Nov: Sihanouk returned as head of state.

1992 Political prisoners released; freedom of speech and party formation restored. Oct: Khmer Rouge refused to disarm in accordance with peace process. Dec: UN Security Council voted to impose limited trade embargo on area of country controlled by Khmer Rouge guerrillas.

1993 General election set for May.

Cameroon
Republic of
(*République du Cameroun*)

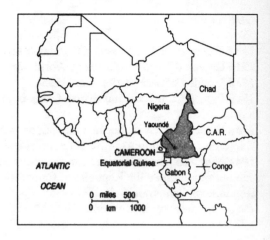

area 475,440 sq km/183,638 sq mi
capital Yaoundé
towns Douala, Nkongsamba, Garova
physical desert in far north in the Lake Chad basin, mountains in W, dry savanna plateau in the intermediate area, and dense tropical rainforest in S
environment the Korup National Park preserves 1,300 sq km/ 500 sq mi of Africa's fast-disappearing tropical rainforest
head of state and of government Paul Biya from 1982
political system emergent democratic republic
exports cocoa, coffee, bananas, cotton, timber, rubber, groundnuts, gold, aluminium, crude oil
currency CFA franc
population (1990 est) 11,109,000; growth rate 2.7% p.a.
life expectancy men 49, women 53
languages French and English in pidgin variations (official); there has been some discontent with the emphasis on French – there are 163 indigenous peoples with their own African languages
religions Roman Catholic 35%, animist 25%, Muslim 22%, Protestant 18%
literacy men 68%, women 45% (1985 est)

GDP $12.7 bn (1987); $1,170 per head
chronology
1884 Treaty signed establishing German rule.
1916 Captured by Allied forces in World War I.
1922 Divided between Britain and France.
1946 French Cameroon and British Cameroons made UN trust territories.
1960 French Cameroon became the independent Republic of Cameroon. Ahmadou Ahidjo elected president.
1961 Northern part of British Cameroon merged with Nigeria and southern part joined the Republic of Cameroon to become the Federal Republic of Cameroon.
1966 One-party regime introduced.
1972 New constitution made Cameroon a unitary state, the United Republic of Cameroon.
1973 New national assembly elected.
1982 Ahidjo resigned and was succeeded by Paul Biya.
1983 Biya began to remove his predecessor's supporters; accused by Ahidjo of trying to create a police state. Ahidjo went into exile in France.
1984 Biya re-elected; defeated a plot to overthrow him. Country's name changed to Republic of Cameroon.
1988 Biya re-elected.
1990 Widespread public disorder. Biya granted amnesty to political prisoners.
1991 Constitutional changes made.
1992 Ruling Democratic Assembly of the Cameroon People (RDPC) won in first multiparty elections in 28 years. Biya's presidential victory challenged by opposition.

Canada

area 9,970,610 sq km/3,849,674 sq mi
capital Ottawa
towns Toronto, Montréal, Vancouver, Edmonton, Calgary, Winnipeg, Quebec
physical mountains in W, with low-lying plains in interior and rolling hills in E. Climate varies from temperate in S to arctic in N
environment sugar maples are dying in E Canada as a result of increasing soil acidification
head of state Elizabeth II from 1952, represented by governor general
head of government Brian Mulroney from 1984
political system federal constitutional monarchy
exports wheat, timber, pulp, newsprint, salmon, furs (ranched fox and mink exceed the value of wild furs), oil, natural gas, aluminium, asbestos, coal, copper, iron, zinc, nickel and uranium (world's largest producer), motor vehicles and parts, fertilizers, chemicals
currency Canadian dollar

population (1990 est) 26,527,000—including 300,000 North American Indians, of whom 75% live on more than 2,000 reservations in Ontario and the four western provinces; some 300,000 Métis (people of mixed race) and 19,000 Inuit of whom 75% live in the Northwest Territories. Growth rate 1.1% p.a.

life expectancy men 72, women 79

languages English, French (both official); there are also North American Indian languages and the Inuit Inuktitut

religion Roman Catholic 46%, Protestant 35%

literacy 99%

GDP $412 bn (1987); $15,910 per head

chronology

1867 Dominion of Canada founded.

1949 Newfoundland joined Canada.

1957 Progressive Conservatives returned to power.

1963 Liberals elected under Lester Pearson.

1968 Pearson succeeded by Pierre Trudeau.

1979 Joe Clark, leader of the Progressive Conservatives, formed a minority government; defeated on budget proposals.

1980 Liberals under Trudeau returned with a large majority. Québec referendum rejected demand for independence.

1982 Canada Act removed Britain's last legal control over Canadian affairs; 'patriation' of Canada's constitution.

1983 Brian Mulroney became leader of the Progressive Conservatives.

1984 Trudeau retired and was succeeded as Liberal leader and prime minister by John Turner. Progressive Conservatives won the federal election with a large majority, and Mulroney became prime minister.

1988 Conservatives re-elected with reduced majority on platform of free trade with the USA.

1990 Collapse of Meech Lake accord. Canada joined the coalition opposing Iraq's invasion of Kuwait.

1991 Constitutional reform package proposed.

1992 Self-governing homeland for Inuit approved. Constitutional reform package, the Charlottetown agreement, rejected in referendum.

1993 Feb: Mulroney resigned leadership of Conservative Party but remained prime minister until a successor was appointed.

Cape Verde
Republic of
(*República de Cabo Verde*)

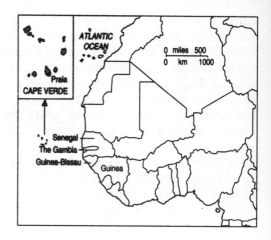

area 4,033 sq km/1,557 sq mi
capital Praia
towns Mindelo, Sal-Rei, Porto Novo
physical archipelago of ten volcanic islands 565 km/350 mi W of Senegal; the windward (Barlavento) group includes Santo Antão, São Vicente, Santa Luzia, São Nicolau, Sal, and Boa Vista; the leeward (Sotovento) group comprises Maio, São Tiago, Fogo, and Brava; all but Santa Luzia are inhabited
head of state Mascarenhas Monteiro from 1991
head of government Carlos Viega from 1991
political system socialist pluralist state
exports bananas, salt, fish
currency Cape Verde escudo
population (1990 est) 375,000 (including 100,000 Angolan refugees); growth rate 1.9% p.a.
life expectancy men 57, women 61
language Creole dialect of Portuguese
religion Roman Catholic 80%
literacy men 61%, women 39% (1985)
GDP $158 million (1987); $454 per head

chronology

15th century First settled by Portuguese.

1951–74 Ruled as an overseas territory by Portugal.

1974 Moved towards independence through a transitional Portuguese–Cape Verde government.

1975 Independence achieved from Portugal. National people's assembly elected. Aristides Pereira became the first president.

1980 Constitution adopted providing for eventual union with Guinea-Bissau.

1981 Union with Guinea-Bissau abandoned and the constitution amended; became one-party state.

1991 First multiparty elections held. New party, Movement for Democracy (MPD), won majority in assembly. Pereira replaced by Mascarenhas Monteiro.

**Central African
Republic**
(*République
Centrafricaine*)

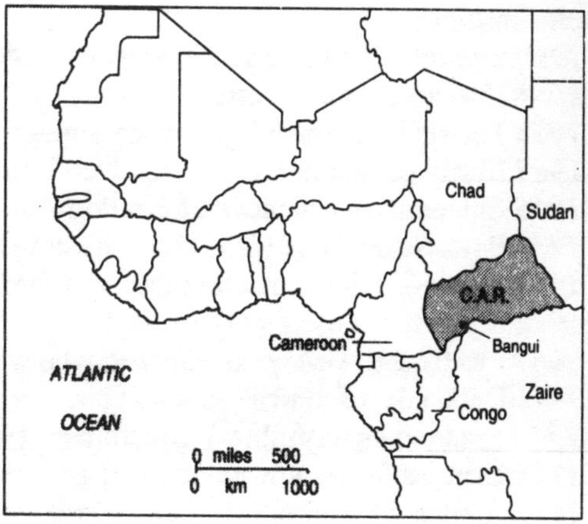

area 622,436 sq km/240,260 sq mi
capital Bangui
towns Berbérati, Bouar, Bossangoa
physical landlocked flat plateau, with rivers flowing N and S, and hills
in NE and SW; dry in N, rainforest in SW
environment an estimated 87% of the urban population is without
access to safe drinking water
head of state and government André Kolingba from 1981
exports diamonds, uranium, coffee, cotton, hardwood timber,
tobacco
currency CFA franc
population (1990 est) 2,879,000 (more than 80 ethnic groups); growth
rate 2.3% p.a.
life expectancy men 41, women 45
languages Sangho (national), French (official), Arabic, Hunsa, and
Swahili
religions Protestant 25%; Roman Catholic 25%; Muslim 10%;
animist 10%
literacy men 53%, women 29% (1985 est)
GDP $1 bn (1987); $374 per head

chronology

1960 Central African Republic achieved independence from France; David Dacko elected president.

1962 The republic made a one-party state.

1965 Dacko ousted in military coup led by Col Bokassa.

1966 Constitution rescinded and national assembly dissolved.

1972 Bokassa declared himself president for life.

1977 Bokassa made himself emperor of the Central African Empire.

1979 Bokassa deposed by Dacko following violent repressive measures by the self-styled emperor, who went into exile.

1981 Dacko deposed in a bloodless coup, led by General André Kolingba, and an all-military government established.

1983 Clandestine opposition movement formed.

1984 Amnesty for all political party leaders announced. President Mitterrand of France paid a state visit.

1985 New constitution promised, with some civilians in the government.

1986 Bokassa returned from France, expecting to return to power; he was imprisoned and his trial started. General Kolingba re-elected. New constitution approved by referendum.

1988 Bokassa found guilty and received death sentence, later commuted to life imprisonment.

1991 Government announced that a national conference would be held in response to demands for a return to democracy.

1992 Abortive debate held on political reform; multiparty elections promised but then postponed.

Chad
Republic of
(*République du
Tchad*)

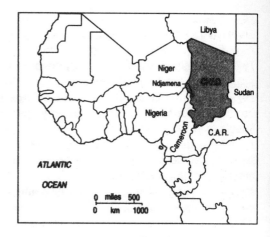

area 1,284,000 sq km/495,624 sq mi
capital Ndjamena (formerly Fort Lamy)
towns Sarh, Moundou, Abéché
physical landlocked state with mountains and part of Sahara Desert in
N; moist savanna in S; rivers in S flow NW to Lake Chad
head of state and government Idriss Deby from 1990
political system emergent democratic republic
exports cotton, meat, livestock, hides, skins
currency CFA franc
population (1990 est) 5,064,000; growth rate 2.3% p.a. Nomadic
tribes move N–S seasonally in search of water
life expectancy men 42, women 45
languages French, Arabic (both official), over 100 African languages
are spoken
religions Muslim 44% (N), Christian 33%, animist 23% (S)
literacy men 40%, women 11% (1985 est)
GDP $980 million (1986); $186 per head
chronology
1960 Independence achieved from France, with François Tombalbaye
as president.

1963 Violent opposition in the Muslim north, led by the Chadian National Liberation Front (Frolinat), backed by Libya.

1968 Revolt quelled with France's help.

1975 Tombalbaye killed in military coup led by Félix Malloum. Frolinat continued its resistance.

1978 Malloum tried to find a political solution by bringing the former Frolinat leader Hissène Habré into his government but they were unable to work together.

1979 Malloum forced to leave the country; an interim government was set up under General Goukouni. Habré continued his opposition with his Army of the North (FAN).

1981 Habré now in control of half the country. Goukouni fled and set up a 'government in exile'.

1983 Habré's regime recognized by the Organization for African Unity (OAU), but in the north Goukouni's supporters, with Libya's help, fought on. Eventually a cease-fire was agreed, with latitude 16°N dividing the country.

1984 Libya and France agreed to a withdrawal of forces.

1985 Fighting between Libyan-backed and French-backed forces intensified.

1987 Chad, France, and Libya agreed on cease-fire proposed by OAU.

1988 Full diplomatic relations with Libya restored.

1989 Libyan troop movements reported on border; Habré re-elected, amended constitution.

1990 President Habré ousted in coup led by Idriss Deby. New constitution adopted.

1991 Several anti-government coups foiled.

1992 Anti-government coup foiled. Two new opposition parties approved.

Chile
Republic of
(*República de Chile*)

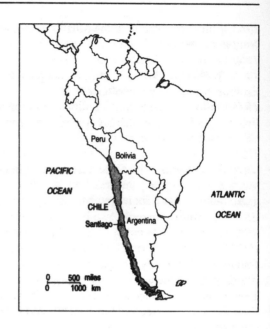

area 756,950 sq km/292,257 sq mi
capital Santiago
towns Concepción, Viña del Mar, Temuco; ports Valparaíso,
Antofagasta, Arica, Iquique, Punta Arenas
physical Andes mountains along E border, Atacama Desert in N,
fertile central valley, grazing land and forest in S
territories Easter Island, Juan Fernández Islands, part of Tierra del
Fuego, claim to part of Antarctica
head of state and government Patricio Aylwin from 1990
political system emergent democratic republic
exports copper (world's leading producer), iron, molybdenum (world's
second largest producer), nitrate, pulp and paper, fishmeal, fruit
currency peso
population (1990 est) 13,000,000 (the majority are of European origin
or are mestizos, of mixed American Indian and Spanish descent);
growth rate 1.6% p.a.

life expectancy men 64, women 73
language Spanish
religion Roman Catholic 89%
literacy 94% (1988)
GDP $18.9 bn (1987); $6,512 per head
chronology
1818 Achieved independence from Spain.
1964 Christian Democratic Party (PDC) formed government under Eduardo Frei.
1970 Dr Salvador Allende became the first democratically elected Marxist president; he embarked on an extensive programme of nationalization and social reform.
1973 Government overthrown by the CIA-backed military, led by General Augusto Pinochet. Allende killed. Policy of repression began during which all opposition was put down and political activity banned.
1983 Growing opposition to the regime from all sides, with outbreaks of violence.
1988 Referendum on whether Pinochet should serve a further term resulted in a clear 'No' vote.
1989 President Pinochet agreed to constitutional changes to allow pluralist politics. Patricio Aylwin (PDC) elected president (his term would begin 1990); Pinochet remained as army commander in chief.
1990 Aylwin reached accord on end to military junta government. Pinochet censured by president.
1992 Future US–Chilean free-trade agreement announced.

China
People's Republic of
(*Zhonghua Renmin
Gonghe Guo*)

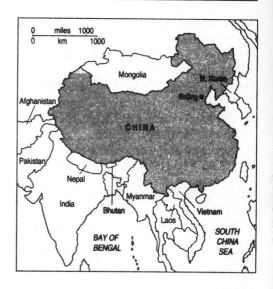

area 9,596,960 sq km/3,599,975 sq mi
capital Beijing (Peking)
towns Chongqing (Chungking), Shenyang (Mukden), Nanjing
(Nanking), Harbin, Tianjin (Tientsin), Shanghai, Guangzhou (Canton)
physical two-thirds of China is mountains or desert (N and W); the
low-lying E is irrigated by rivers Huang He (Yellow River), Chang
Jiang (Yangtze-Kiang), Xi Jiang (Si Kiang)
head of state Yang Shangkun from 1988
head of government Li Peng from 1987
political system communist republic
exports tea, livestock and animal products, silk, cotton, oil, minerals
(China is the world's largest producer of tungsten and antimony),
chemicals, light industrial goods
currency yuan
population (1990 est) 1,130,065,000 (the majority are Han or ethnic
Chinese; the 67 million of other ethnic groups, including Tibetan,
Uigur, and Zhuang, live in border areas). The number of people of
Chinese origin outside China, Taiwan, and Hong Kong is estimated at

15–24 million. Growth rate 1.2% p.a.
life expectancy men 67, women 69
languages Chinese, including Mandarin (official), Cantonese, and other dialects
religions officially atheist, but traditionally Taoist, Confucianist, and Buddhist; Muslim 13 million; Catholic 3–6 million (divided between the 'patriotic' church established 1958 and the 'loyal' church subject to Rome); Protestant 3 million
literacy men 82%, women 66% (1985 est)
GDP $293.4 bn (1987); $274 per head
chronology
1949 People's Republic of China proclaimed by Mao Zedong.
1954 Soviet-style constitution adopted.
1956–57 Hundred Flowers Movement encouraged criticism of the government.
1958–60 Great Leap Forward commune experiment to achieve 'true communism'.
1960 Withdrawal of Soviet technical advisers.
1962 Sino-Indian border war.
1962–65 Economic recovery programme under Liu Shaoqi; Maoist 'socialist education movement' rectification campaign.
1966–69 Great Proletarian Cultural Revolution; Liu Shaoqi overthrown.
1969 Ussuri River border clashes with USSR.
1970–76 Reconstruction under Mao and Zhou Enlai.
1971 Entry into United Nations.
1972 US president Nixon visited Beijing.
1975 New state constitution. Unveiling of Zhou's 'Four Modernizations' programme.
1976 Deaths of Zhou Enlai and Mao Zedong; appointment of Hua Guofeng as prime minister and Communist Party chair. Vice Premier Deng Xiaoping in hiding. Gang of Four arrested.
1977 Rehabilitation of Deng Xiaoping.
1979 Economic reforms introduced. Diplomatic relations opened with USA. Punitive invasion of Vietnam.
1980 Zhao Ziyang appointed prime minister.

1981 Hu Yaobang succeeded Hua Guofeng as party chair.
Imprisonment of Gang of Four.

1982 New state constitution adopted.

1984 'Enterprise management' reforms for industrial sector.

1986 Student prodemocracy demonstrations.

1987 Hu was replaced as party leader by Zhao, with Li Peng as prime
minister. Deng left Politburo but remained influential.

1988 Yang Shangkun replaced Li Xiannian as state president.
Economic reforms encountered increasing problems; inflation
rocketed.

1989 Over 2,000 killed in prodemocracy student demonstrations in
Tiananmen Square; international sanctions imposed.

1991 March: European Community and Japanese sanctions lifted.
May: normal relations with USSR resumed. Sept: UK prime minister
John Major visited Beijing. Nov: relations with Vietnam normalized.

1992 China promised to sign 1968 Nuclear Non-Proliferation Treaty.
Historic visit by Japan's emperor.

1993 Jiang Zemin, Chinese Communist Party general secretary, set to
replace Yang Shangkun as president.

Colombia
Republic of
(*República de
Colombia*)

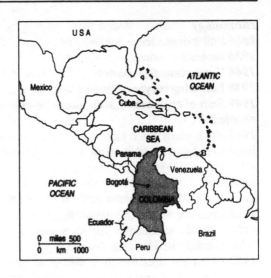

area 1,141,748 sq km/440,715 sq mi
capital Bogotá
towns Medellín, Cali, Bucaramanga, Barranquilla, Cartagena,
physical the Andes mountains run N–S; flat coastland in W and plains
(llanos) in E; Magdalena River runs N to Caribbean Sea; includes
islands of Providencia, San Andrés, and Mapelo
head of state and government Cesar Gaviria Trujillo from 1990
political system emergent democratic republic
exports emeralds (world's largest producer), coffee (world's second
largest producer), cocaine (country's largest export), bananas, cotton,
meat, sugar, oil, skins, hides, tobacco
currency peso
population (1990 est) 32,598,800 (mestizo 68%, white 20%,
Amerindian 1%); growth rate 2.2% p.a.
life expectancy men 61, women 66; Indians 34
language Spanish
religion Roman Catholic 95%
literacy men 89%, women 87% (1987); Indians 40%
GDP $31.9 bn (1987); $1,074 per head

chronology
1886 Full independence achieved from Spain. Conservatives in power.
1930 Liberals in power.
1946 Conservatives in power.
1948 Left-wing mayor of Bogotá assassinated; widespread outcry.
1949 Start of civil war, 'La Violencia', during which over 250,000 people died.
1957 Hoping to halt the violence, Conservatives and Liberals agreed to form a National Front, sharing the presidency.
1970 National Popular Alliance (ANAPO) formed as a left-wing opposition to the National Front.
1974 National Front accord temporarily ended.
1975 Civil unrest because of disillusionment with the government.
1978 Liberals, under Julio Turbay, revived the accord and began an intensive fight against drug dealers.
1982 Liberals maintained their control of congress but lost the presidency. The Conservative president, Belisario Betancur, granted guerrillas an amnesty and freed political prisoners.
1984 Minister of justice assassinated by drug dealers; campaign against them stepped up.
1986 Virgilio Barco Vargas, Liberal, elected president by record margin.
1989 Drug cartel assassinated leading presidential candidate; Vargas declared antidrug war; bombing campaign by drug lords killed hundreds; police killed José Rodríguez Gacha, one of the most wanted cartel leaders.
1990 Cesar Gaviria Trujillo elected president. Liberals maintained control of congress.
1991 New constitution prohibited extradition of Colombians wanted for trial in other countries; several leading drug traffickers arrested. Oct: Liberal Party won general election.
1992 One of leading drug barons, Pablo Escobar, escaped from prison.
1993 Escobar continued to defy government.

Comoros
Federal Islamic
Republic of
(*Jumhurīyat al-
Qumur al-
Itthādīyah al-
Islāmīyah*)

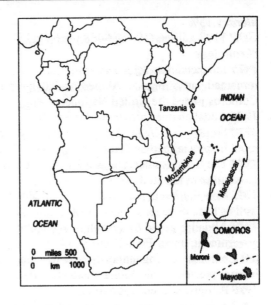

area 1,862 sq km/719 sq mi
capital Moroni
towns Mutsamudu, Domoni, Fomboni
physical comprises the volcanic islands of Njazídja, Nzwani, and
Mwali (formerly Grande Comore, Anjouan, Moheli); at N end of
Mozambique Channel
head of state Said Mohammad Djohar (interim administration)
from 1989
head of government Halidi Abderamane Ibrahim from 1993
political system authoritarian nationalism
exports copra, vanilla, cocoa, sisal, coffee, cloves, essential oils
currency CFA franc
population (1990 est) 459,000; growth rate 3.1% p.a.
life expectancy men 48, women 52
languages Arabic (official), Comorian (Swahili and Arabic dialect),
Makua, French
religions Muslim (official) 86%, Roman Catholic 14%

literacy 15%

GDP $198 million (1987); $468 per head

chronology

1975 Independence achieved from France, but island of Mayotte remained part of France. Ahmed Abdallah elected president. The Comoros joined the United Nations.

1976 Abdallah overthrown by Ali Soilih.

1978 Soilih killed by mercenaries working for Abdallah. Islamic republic proclaimed and Abdallah elected president.

1979 The Comoros became a one-party state; powers of the federal government increased.

1985 Constitution amended to make Abdallah head of government as well as head of state.

1989 Abdallah killed by French mercenaries who took control of government; under French and South African pressure, mercenaries left Comoros, turning authority over to French administration and interim president Said Mohammad Djohar.

1990 Antigovernment coup foiled.

1992 Third transitional government appointed. Antigovernment coup foiled.

1993 Jan: general election failed to provide any one party with overall assembly majority. President Djohar appointed Halidi Abderamane Ibrahim prime minister.

Congo
Republic of
(*République du Congo*)

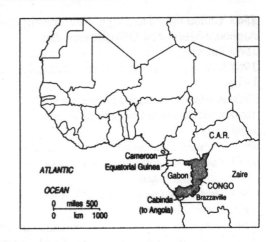

area 342,000 sq km/132,012 sq mi
capital Brazzaville
towns Pointe-Noire, N'Kayi, Loubomo
physical narrow coastal plain rises to central plateau, then falls into northern basin; Zaïre (Congo) River on the border with Zaire; half the country is rainforest
environment an estimated 93% of the rural population is without access to safe drinking water
head of state and government Pascal Lissouba from 1992
political system emergent democracy
exports timber, petroleum, cocoa, sugar
currency CFA franc
population (1990 est) 2,305,000 (chiefly Bantu); growth rate 2.6% p.a.
life expectancy men 45, women 48
languages French (official); many African languages
religions animist 50%, Christian 48%, Muslim 2%
literacy men 79%, women 55% (1985 est)
GDP $2.1 bn (1983); $500 per head
chronology
1910 Became part of French Equatorial Africa.
1960 Achieved independence, with Abbé Youlou as first president.

1963 Youlou forced to resign. New constitution approved, with Alphonse Massamba-Débat as president.

1964 The Congo became a one-party state.

1968 Military coup, led by Capt Marien Ngouabi, ousted Massamba-Débat.

1970 A Marxist state, the People's Republic of the Congo, was announced, with the Congolese Labour Party (PCT) as the only legal party.

1977 Ngouabi assassinated. Col Yhombi-Opango became president.

1979 Yhombi-Opango handed over the presidency to the PCT, who chose Col Denis Sassou-Nguessou as his successor.

1984 Sassou-Nguessou elected for another five-year term.

1990 The PCT abandoned Marxist-Leninism and promised multiparty politics.

1991 1979 constitution suspended. Country renamed the Republic of Congo.

1992 New constitution approved and multiparty elections held, giving Pan-African Union for Social Democracy (UPADS) the most assembly seats.

Costa Rica
Republic of
(*República de
Costa Rica*)

area 51,100 sq km/19,735 sq mi
capital San José
towns Limón, Puntarenas
physical high central plateau and tropical coasts; Costa Rica was once
entirely forested, containing an estimated 5% of the Earth's flora and
fauna. By 1983 only 17% of the forest remained; half of the arable
land had been cleared for cattle ranching, which led to landlessness,
unemployment (except for 2,000 politically powerful families), and
soil erosion; the massive environmental destruction also caused
incalculable loss to the gene pool
environment one of the leading centres of conservation in Latin
America, with more than 10% of the country protected by national
parks, and tree replanting proceeding at a rate of 150 sq km/60 sq mi
per year
head of state and government Rafael Calderón from 1990
political system liberal democracy
exports coffee, bananas, cocoa, sugar, beef
currency colón
population (1990 est) 3,032,000 (including 1,200 Guaymi Indians);
growth rate 2.6% p.a.

life expectancy men 71, women 76
language Spanish (official)
religion Roman Catholic 95%
literacy men 94%, women 93% (1985 est)
GDP $4.3 bn (1986); $1,550 per head
chronology
1821 Independence achieved from Spain.
1949 New constitution adopted. National army abolished. José
Figueres, cofounder of the National Liberation Party (PLN) elected
president; he embarked on ambitious socialist programme.
1958–73 Mainly conservative administrations.
1974 PLN regained the presidency and returned to socialist policies.
1978 Rodrigo Carazo, conservative, elected president. Sharp
deterioration in the state of the economy.
1982 Luis Alberto Monge (PLN) elected president. Harsh austerity
programme introduced to rebuild the economy. Pressure from the USA
to abandon neutral stance and condemn Sandinista regime in
Nicaragua.
1983 Policy of neutrality reaffirmed.
1985 Following border clashes with Sandinista forces, a US-trained
antiguerrilla guard formed.
1986 Oscar Arias won the presidency on a neutralist platform.
1987 Arias won Nobel Prize for Peace for devising a Central
American peace plan.
1990 Rafael Calderón, Christian Socialist Union Party (PUSC),
elected president.

Croatia
Republic of

area 56,538 sq km/21,824 sq mi
capital Zagreb
towns Rijeka (Fiume), Zadar, Sibenik, Split, Dubrovnik
physical Adriatic coastline with large islands; very mountainous, with part of the Karst region and the Julian and Styrian Alps
head of state Franjo Tudjman from 1990
head of government Hrvoje Sarinic from 1992
political system emergent democracy
products cereals, potatoes, tobacco, fruit, livestock, metal goods, textiles
currency Croatian dinar
population (1990) 4,760,000 including 75% Croats, 12% Serbs, and 1% Slovenes
language Croatian variant of Serbo-Croatian
religions Roman Catholic (Croats); Orthodox Christian (Serbs)
GNP $7.9 bn (1990); $1,660 per head
chronology
1918 Became part of the kingdom that united the Serbs, Croats,

and Slovenes.

1929 The kingdom of Croatia, Serbia, and Slovenia became Yugoslavia. Croatia continued its campaign for autonomy.

1941 Became a Nazi puppet state following German invasion.

1945 Became constituent republic of Yugoslavia.

1970s Separatist demands resurfaced. Crackdown against anti-Serb separatist agitators.

1989 Formation of opposition parties permitted.

1990 April–May: Communists defeated by Tudjman-led Croatian Democratic Union (HDZ) in first free election since 1938. Sept: 'sovereignty' declared. Dec: new constitution adopted.

1991 Feb: assembly called for Croatia's secession. March: Serb-dominated Krajina announced secession from Croatia. June: Croatia declared independence; military conflict with Serbia; internal civil war ensued. July onwards: civil war intensified. Oct: Croatia formally seceded from Yugoslavia.

1992 Jan: United Nations peace accord reached in Sarajevo; Croatia's independence recognized by the European Community. March–April: UN peacekeeping forces drafted into Croatia. April: independence recognized by USA. May: became a member of the United Nations. Aug: Tudjman directly elected president; HDZ won assembly elections. Sept: Tudjman requested withdrawal of UN forces on expiry of mandate 1993.

1993 Jan: Croatian forces launched offensive to retake parts of Serb-held Krajina, violating the 1992 UN peace accord.

Cuba
Republic of
(*República de Cuba*)

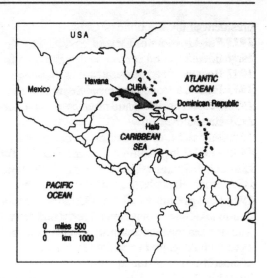

area 110,860 sq km/42,820 sq mi
capital Havana
towns Santiago de Cuba, Camagüey
physical comprises Cuba, the largest and westernmost of the West
Indies, and smaller islands including Isle of Youth; low hills; Sierra
Maestra mountains in SE
head of state and government Fidel Castro Ruz from 1959
political system communist republic
exports sugar, tobacco, coffee, nickel, fish
currency Cuban peso
population (1990 est) 10,582,000; 37% are white of Spanish descent,
51% mulatto, and 11% are of African origin; growth rate 0.6% p.a.
life expectancy men 72, women 75
language Spanish
religions Roman Catholic 85%; also Episcopalians and Methodists
literacy men 96%, women 95% (1988)
disposable national income $15.8 bn (1983); $1,590 per head
chronology
1901 Cuba achieved independence; Tomás Estrada Palma became first

president of the Republic of Cuba.

1933 Fulgencia Batista seized power.

1944 Batista retired.

1952 Batista seized power again to begin an oppressive regime.

1953 Fidel Castro led an unsuccessful coup against Batista.

1956 Second unsuccessful coup by Castro.

1959 Batista overthrown by Castro. Constitution of 1940 replaced by a 'Fundamental Law', making Castro prime minister, his brother Raúl Castro his deputy, and 'Che' Guevara his number three.

1960 All US businesses in Cuba appropriated without compensation; USA broke off diplomatic relations.

1961 USA sponsored an unsuccessful invasion at the Bay of Pigs. Castro announced that Cuba had become a communist state, with a Marxist-Leninist programme of economic development.

1962 Cuba expelled from the Organization of American States. Soviet nuclear missiles installed but subsequently removed from Cuba at US insistence.

1965 Cuba's sole political party renamed Cuban Communist Party (PCC). With Soviet help, Cuba began to make considerable economic and social progress.

1972 Cuba became a full member of the Moscow-based Council for Mutual Economic Assistance.

1976 New socialist constitution approved; Castro elected president.

1976–81 Castro became involved in extensive international commitments, sending troops as Soviet surrogates, particularly to Africa.

1982 Cuba joined other Latin American countries in giving moral support to Argentina in its dispute with Britain over the Falklands.

1984 Castro tried to improve US-Cuban relations by discussing exchange of US prisoners in Cuba for Cuban 'undesirables' in the USA.

1988 Peace accord with South Africa signed, agreeing to withdrawal of Cuban troops from Angola.

1989 Reduction in Cuba's overseas military activities.

1991 Soviet troops withdrawn.

1992 Castro affirmed continuing support of communism.

Cyprus
Greek *Republic
of Cyprus*
(*Kypriakí
Dimokratía*)
in the south, and
*Turkish Republic of
Northern Cyprus*
(*Kibris Cumhuriyeti*)
in the north

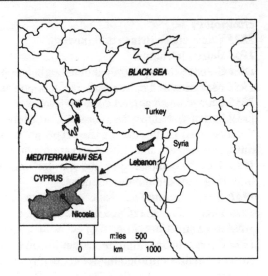

area 9,251 sq km/3,571 sq mi, 37% in Turkish hands
capital Nicosia (divided between Greeks and Turks)
towns Limassol, Larnaca, Paphos (Greek); Morphou, Kyrenia,
Famagusta (Turkish)
physical central plain between two E–W mountain ranges
heads of state and government Glafkos Clerides (Greek) from 1993,
Rauf Denktaş (Turkish) from 1976
political system democratic divided republic
exports citrus, grapes, raisins, Cyprus sherry, potatoes, clothing,
footwear
currency Cyprus pound and Turkish lira
population (1990 est) 708,000 (Greek Cypriot 78%, Turkish Cypriot
18%); growth rate 1.2% p.a.
life expectancy men 72, women 76
languages Greek and Turkish (official), English
religions Greek Orthodox 78%, Sunni Muslim 18%
literacy 99% (1984)
GDP $3.7 bn (1987); $5,497 per head

chronology

1878 Came under British administration.

1955 Guerrilla campaign began against the British for enosis (union with Greece), led by Archbishop Makarios and General Grivas.

1956 Makarios and enosis leaders deported.

1959 Compromise agreed and Makarios returned to be elected president of an independent Greek–Turkish Cyprus.

1960 Independence achieved from Britain, with Britain retaining its military bases.

1963 Turks set up their own government in northern Cyprus. Fighting broke out between the two communities.

1964 United Nations peacekeeping force installed.

1971 Grivas returned to start a guerrilla war against the Makarios government.

1974 Grivas died. Military coup deposed Makarios, who fled to Britain. Nicos Sampson appointed president. Turkish army sent to northern Cyprus to confirm Turkish Cypriots' control; military regime in southern Cyprus collapsed; Makarios returned. Northern Cyprus declared itself the Turkish Federated State of Cyprus (TFSC), with Rauf Denktaş as president.

1977 Makarios died; succeeded by Spyros Kyprianou.

1983 An independent Turkish Republic of Northern Cyprus proclaimed but recognized only by Turkey.

1984 UN peace proposals rejected.

1985 Summit meeting between Kyprianou and Denktaş failed to reach agreement.

1988 Georgios Vassiliou elected president. Talks with Denktaş began, under UN auspices.

1989 Vassiliou and Denktaş agreed to draft an agreement for the future reunification of the island, but peace talks were abandoned Sept.

1991 Turkish offer of peace talks rejected by Cyprus and Greece.

1992 UN-sponsored peace talks collapsed.

1993 Glafkos Clerides replaced Vassiliou as Greek president.

Czech Republic
(*Ceská Republika*)

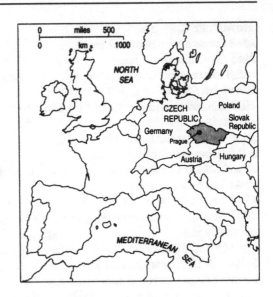

area 78,864 sq km/30,461 sq mi
capital Prague
towns Brno, Ostrava, Olomouc, Liberec, Plzen, Ustí nad Labem
physical mountainous; rivers: Morava, Labe (Elbe), Vltava (Moldau)
environment considered in 1991 to be the most polluted country in
E Europe. Pollution is worst in N Bohemia, which produces 70% of
the country's coal and 45% of its coal-generated electricity. Up to 20
times the permissible level of sulphur dioxide is released over Prague,
where 75% of the drinking water fails to meet the country's health
standards
head of state Václav Havel from 1993
head of government Václav Klaus from 1993
political system emergent democracy
exports machinery, vehicles, coal, iron and steel, chemicals, glass,
ceramics, clothing
currency new currency based on koruna
population (1991) 10,298,700 (with German and other minorities);

growth rate 0.4% p.a.
life expectancy men 68, women 75
languages Czech (official)
religions Roman Catholic 75%, Protestant, Hussite, Orthodox
literacy 100%
GDP $26,600 million (1990); $2,562 per head
chronology
1526–1918 Under Habsburg domination.
1918 Independence achieved from Austro-Hungarian Empire; Czechs joined Slovaks in forming Czechoslovakia as independent nation.
1948 Communists assumed power in Czechoslovakia.
1968 Czech Socialist Republic created under new federal constitution.
1989 Nov: pro-democracy demonstrations in Prague; new political parties formed, including Czech-based Civic Forum under Václav Havel; Communist Party stripped of powers; political parties legalized. Dec: new 'grand coalition' government formed, including former dissidents; Havel appointed state president. Amnesty granted to 22,000 prisoners; calls for USSR to withdraw troops.
1990 July: Havel re-elected president in multiparty elections.
1991 Civic Forum split into Civil Democratic Party (CDP) and Civic Movement (CM); evidence of increasing Czech and Slovak separatism.
1992 June: Václav Klaus, leader of the Czech-based CDP, became prime minister; Havel resigned following Slovak gains in assembly elections. Aug: creation of separate Czech and Slovak states agreed.
1993 Jan: Czech Republic became sovereign state, with Klaus as prime minister. Havel elected president of the new republic. Admitted into United Nations, Conference on Security and Cooperation in Europe, and Council of Europe.

Denmark
Kingdom of
(*Kongeriget
Danmark*)

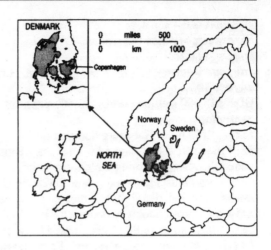

area 43,075 sq km/16,627 sq mi
capital Copenhagen
towns Aarhus, Odense, Aalborg, Esbjerg
physical comprises the Jutland peninsula and about 500 islands (100 inhabited) including Bornholm in the Baltic Sea; the land is flat and cultivated; sand dunes and lagoons on the W coast and long inlets (fjords) on the E; the main island is Sjælland (Zealand), where most of Copenhagen is located (the rest is on the island of Amager)
territories the dependencies of Faeroe Islands and Greenland
head of state Queen Margrethe II from 1972
head of government Poul Nyrup Rasmussen from 1993
political system liberal democracy
exports bacon, dairy produce, eggs, fish, mink pelts, car and aircraft parts, electrical equipment, textiles, chemicals
currency kroner
population (1990 est) 5,134,000; growth rate 0% p.a.
life expectancy men 72, women 78
languages Danish (official); there is a German-speaking minority
religion Lutheran 97%
literacy 99% (1983)
GDP $85.5 bn (1987); $16,673 per head

recent chronology

1940–45 Occupied by Germany.

1945 Iceland's independence recognized.

1947 Frederik IX succeeded Christian X.

1948 Home rule granted for Faeroe Islands.

1949 Became a founding member of NATO.

1960 Joined European Free Trade Association (EFTA).

1972 Margrethe II became Denmark's first queen in nearly 600 years.

1973 Left EFTA and joined European Economic Community (EEC).

1979 Home rule granted for Greenland.

1985 Strong non-nuclear movement in evidence.

1990 General election; another coalition government formed.

1992 Rejection of Maastricht Treaty in national referendum; government requested modifications (codicils) to treaty prior to second national referendum, planned for 1993.

1993 Poul Schlüter resigned; replaced by Poul Nyrup Rasmussen at head of Social Democrat-led coalition government.

Djibouti
Republic of
(*Jumhouriyya
Djibouti*)

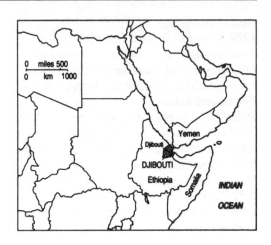

area 23,200 sq km/8,955 sq mi
capital (and chief port) Djibouti
towns Tadjoura, Obock, Dikhil
physical mountains divide an inland plateau from a coastal plain; hot and arid
head of state and government Hassan Gouled Aptidon from 1977
political system authoritarian nationalism
exports acts mainly as a transit port for Ethiopia
currency Djibouti franc
population (1990 est) 337,000 (Issa 47%, Afar 37%, European 8%, Arab 6%); growth rate 3.4% p.a.
life expectancy 50
languages French (official), Somali, Afar, Arabic
religion Sunni Muslim
literacy 20% (1988)
GDP $378 million (1987); $1,016 per head
chronology
1884 Annexed by France as part of French Somaliland.
1967 French Somaliland became the French Territory of the Afars and the Issas.
1977 Independence achieved from France; Hassan Gouled was

elected president.

1979 All political parties combined to form the People's Progress Assembly (RPP).

1981 New constitution made RPP the only legal party. Gouled re-elected. Treaties of friendship signed with Ethiopia, Somalia, Kenya, and Sudan.

1984 Policy of neutrality reaffirmed.

1987 Gouled re-elected for a final term.

1991 Amnesty International accused secret police of brutality.

1992 Djibouti elected member of UN Security Council 1993–95.

Dominica
Commonwealth of

area 751 sq km/290 sq mi
capital Roseau, with a deepwater port
towns Portsmouth, Marigot
physical second largest of the Windward Islands, mountainous central ridge with tropical rainforest
head of state Clarence Seignoret from 1983
head of government Eugenia Charles from 1980
political system liberal democracy
exports bananas, coconuts, citrus, lime, bay oil
currency Eastern Caribbean dollar, pound sterling, French franc
population (1990 est) 94,200 (mainly black African in origin, but with a small Carib reserve of some 500); growth rate 1.3% p.a.
life expectancy men 57, women 59
language English (official), but the Dominican patois reflects earlier periods of French rule
religion Roman Catholic 80%
literacy 80%
GDP $91 million (1985); $1,090 per head
chronology
1763 Became British possession.

1978 Independence achieved from Britain. Patrick John, leader of Dominica Labour Party (DLP), elected prime minister.

1980 Dominica Freedom Party (DFP), led by Eugenia Charles, won convincing victory in general election.

1981 Patrick John implicated in plot to overthrow government.

1982 John tried and acquitted.

1985 John retried and found guilty. Regrouping of left-of-centre parties resulted in new Labour Party of Dominica (LPD). DFP, led by Eugenia Charles, re-elected.

1990 Charles elected to a third term.

1991 Integration into Windward Islands confederation proposed.

Dominican Republic
(*República Dominicana*)

area 48,442 sq km/18,700 sq mi
capital Santo Domingo
towns Santiago de los Caballeros, San Pedro de Macoris
physical comprises eastern two-thirds of island of Hispaniola; central mountain range with fertile valleys
head of state and government Joaquín Ricardo Balaguer from 1986
political system democratic republic
exports sugar, gold, silver, tobacco, coffee, nickel
currency peso
population (1989 est) 7,307,000; growth rate 2.3% p.a.
life expectancy men 61, women 65
language Spanish (official)
religion Roman Catholic 95%
literacy men 78%, women 77% (1985 est)
GDP $4.9 bn (1987); $731 per head
chronology
1844 Dominican Republic established.
1930 Military coup established dictatorship of Rafael Trujillo.
1961 Trujillo assassinated.
1962 First democratic elections resulted in Juan Bosch, founder of the

Dominican Revolutionary Party (PRD), becoming president.

1963 Bosch overthrown in military coup.

1965 US Marines intervene to restore order and protect foreign nationals.

1966 New constitution adopted. Joaquín Balaguer, leader of Christian Social Reform Party (PRSC), became president.

1978 PRD returned to power, with Silvestre Antonio Guzmán as president.

1982 PRD re-elected, with Jorge Blanco as president.

1985 Blanco forced by International Monetary Fund to adopt austerity measures to save the economy.

1986 PRSC returned to power, with Balaguer re-elected president.

Ecuador
Republic of
(*República del
Ecuador*)

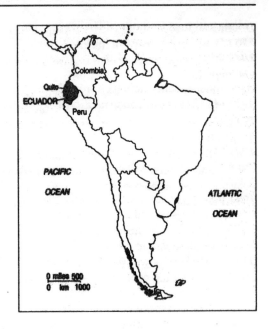

area 270,670 sq km/104,479 sq mi
capital Quito
towns Cuenca, Guayaquil
physical coastal plain rises sharply to Andes Mountains, which are
divided into a series of cultivated valleys; low-lying rainforest in E
environment about 25,000 species became extinct 1965–90 as a result
of environmental destruction
head of state and government Sixto Duran Ballen from 1992
political system emergent democracy
exports bananas, cocoa, coffee, sugar, rice, fruit, fish, petroleum
currency sucre
population (1989 est) 10,490,000; (mestizo 55%, Indian 25%,
European 10%, black African 10%); growth rate 2.9% p.a.
life expectancy men 62, women 66
languages Spanish (official); Quechua, Jivaro, and other Indian
languages

religion Roman Catholic 95%
literacy men 85%, women 80% (1985 est)
GDP $10.6 bn (1987); $1,069 per head
chronology
1830 Independence achieved from Spain.
1925–48 Great political instability; no president completed his term of office.
1948–55 Liberals in power.
1956 First conservative president in 60 years.
1960 Liberals returned, with José Velasco as president.
1961 Velasco deposed and replaced by the vice president.
1962 Military junta installed.
1968 Velasco returned as president.
1972 A coup put the military back in power.
1978 New democratic constitution adopted.
1979 Liberals in power but opposed by right- and left-wing parties.
1982 Deteriorating economy provoked strikes, demonstrations, and a state of emergency.
1983 Austerity measures introduced.
1984–85 No party with a clear majority in the national congress; Febres Cordero narrowly won the presidency for the Conservatives.
1988 Rodrigo Borja Cevallos elected president for moderate left-wing coalition.
1989 Guerrilla left-wing group, *Alfaro Vive, Carajo* ('Alfaro lives, Dammit'), numbering about 1,000, laid down arms after nine years.
1992 United Republican Party (PUR) leader, Sixto Duran Ballen, elected president; Social Christian Party (PSC) became largest party in congress.

Egypt
Arab Republic of
(*Jumhuriyat Misr al-Arabiya*)

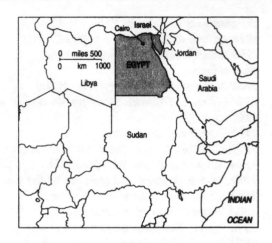

area 1,001,450 sq km/386,990 sq mi
capital Cairo
towns Gîza, Alexandria, Port Said, Suez, Damietta
physical mostly desert; hills in E; fertile land along Nile valley and delta; cultivated and settled area is about 35,500 sq km/13,700 sq mi
environment the building of the Aswan Dam (opened 1970) on the Nile has caused widespread salinization and an increase in waterborne diseases in villages close to Lake Nasser. A dramatic fall in the annual load of silt deposited downstream has reduced the fertility of cropland and has led to coastal erosion and the consequent loss of sardine shoals
head of state and government Hosni Mubarak from 1981
political system democratic republic
exports cotton and textiles, petroleum, fruit and vegetables
currency Egyptian pound
population (1989 est) 54,779,000; growth rate 2.4% p.a.
life expectancy men 57, women 60
languages Arabic (official); ancient Egyptian survives to some extent in Coptic
religions Sunni Muslim 95%, Coptic Christian 5%
literacy men 59%, women 30% (1985 est)
GDP $34.5 bn (1987); $679 per head

chronology

1914 Egypt became a British protectorate.

1936 Independence achieved from Britain. King Fuad succeeded by his son Farouk.

1946 Withdrawal of British troops except from Suez Canal Zone.

1952 Farouk overthrown by army in bloodless coup.

1953 Egypt declared a republic, with General Neguib as president.

1956 Neguib replaced by Col Gamal Nasser. Nasser announced nationalization of Suez Canal; Egypt attacked by Britain, France, and Israel. Cease-fire agreed because of US intervention.

1958 Short-lived merger of Egypt and Syria as United Arab Republic (UAR). Subsequent attempts to federate Egypt, Syria, and Iraq failed.

1967 Six-Day War with Israel ended in Egypt's defeat and Israeli occupation of Sinai and Gaza Strip.

1970 Nasser died suddenly; succeeded by Anwar Sadat.

1973 Attempt to regain territory lost to Israel led to fighting; cease-fire arranged by US secretary of state Henry Kissinger.

1977 Sadat's visit to Israel to address the Israeli parliament was criticized by Egypt's Arab neighbours.

1978–79 Camp David talks in the USA resulted in a treaty between Egypt and Israel. Egypt expelled from the Arab League.

1981 Sadat assassinated, succeeded by Hosni Mubarak.

1983 Improved relations between Egypt and the Arab world; only Libya and Syria maintained a trade boycott.

1984 Mubarak's party victorious in the people's assembly elections.

1987 Mubarak re-elected. Egypt readmitted to Arab League.

1988 Full diplomatic relations with Algeria restored.

1989 Improved relations with Libya; diplomatic relations with Syria restored. Mubarak proposed a peace plan.

1991 Participation in Gulf War on US-led side. Major force in convening Middle East peace conference in Spain.

1992 Outbreaks of violence between Muslims and Christians. Earthquake devastated Cairo.

El Salvador
Republic of
(*República de El Salvador*)

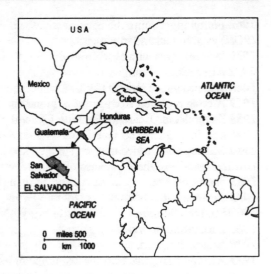

area 21,393 sq km/8,258 sq mi
capital San Salvador
towns Santa Ana, San Miguel
physical narrow coastal plain, rising to mountains in N with central plateau
head of state and government Alfredo Cristiani from 1989
political system emergent democracy
exports coffee, cotton, sugar
currency colón
population (1989 est) 5,900,000 (mainly of mixed Spanish and Indian ancestry; 10% Indian); growth rate 2.9% p.a.
life expectancy men 63, women 66
languages Spanish, Nahuatl
religion Roman Catholic 97%
literacy men 75%, women 69% (1985 est)
GDP $4.7 bn (1987); $790 per head
chronology
1821 Independence achieved from Spain.
1931 Peasant unrest followed by a military coup.

1961 Following a coup, the conservative National Conciliation Party (PCN) established and in power.

1969 'Soccer' war with Honduras.

1972 Allegations of human-rights violations; growth of left-wing guerrilla activities. General Carlos Romero elected president.

1979 A coup replaced Romero with a military–civilian junta.

1980 Archbishop Oscar Romero assassinated; country on verge of civil war. José Duarte became first civilian president since 1931.

1981 Mexico and France recognized the guerrillas as a legitimate political force, but the USA actively assisted the government in its battle against them.

1982 Assembly elections boycotted by left-wing parties and held amid considerable violence.

1986 Duarte sought a negotiated settlement with the guerrillas.

1988 Duarte resigned.

1989 Alfredo Cristiani, National Republican Alliance (ARENA), became president in rigged elections; rebel attacks intensified.

1991 UN-sponsored peace accord signed by representatives of the government and the socialist guerrilla group, the Farabundo Marti Liberation Front (FMLN).

1992 Peace accord validated; FMLN became political party.

Equatorial Guinea
Republic of
(*República de
Guinea Ecuatorial*)

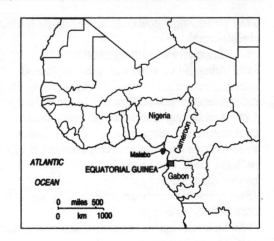

area 28,051 sq km/10,828 sq mi
capital Malabo (Bioko)
towns Bata, Mbini (Río Muni)
physical comprises mainland Río Muni, plus the small islands of
Corisco, Elobey Grande and Elobey Chico, and Bioko (formerly
Fernando Po) together with Annobón (formerly Pagalu)
head of state and government Teodoro Obiang Nguema Mbasogo
from 1979
political system one-party military republic
exports cocoa, coffee, timber
currency ekuele; CFA franc
population (1988 est) 336,000 (plus 110,000 estimated to live in exile
abroad); growth rate 2.2% p.a.
life expectancy men 44, women 48
languages Spanish (official); pidgin English is widely spoken, and on
Annobón (whose people were formerly slaves of the Portuguese) a
Portuguese dialect; Fang and other African dialects spoken on Río
Muni
religions nominally Christian, mainly Catholic, but in 1978 Roman
Catholicism was banned
literacy 55% (1984)

GDP $90 million (1987); $220 per head
chronology
1778 Fernando Po (Bioko Island) ceded to Spain.
1885 Mainland territory came under Spanish rule; colony known as Spanish Guinea.
1968 Independence achieved from Spain. Francisco Macias Nguema became first president, soon assuming dictatorial powers.
1979 Macias overthrown and replaced by his nephew, Teodoro Obiang Nguema Mbasogo, who established a military regime. Macias tried and executed.
1982 Obiang elected president unopposed for another seven years. New constitution adopted.
1989 Obiang re-elected president.
1992 New constitution adopted; elections held, but president continued to nominate candidates for top government posts.

Estonia
Republic of

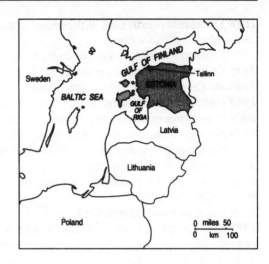

area 45,000 sq km/17,000 sq mi
capital Tallinn
towns Tartu, Narva, Kohtla-Järve, Pärnu
physical lakes and marshes in a partly forested plain; 774 km/481 mi
of coastline; mild climate
head of state Lennart Meri from 1992
head of government Tiit Vahl from 1992
political system emergent democracy
products oil and gas (from shale), wood products, flax, dairy and
pig products
currency kroon
population (1989) 1,573,000 (Estonian 62%, Russian 30%, Ukrainian
3%, Byelorussian 2%)
language Estonian, allied to Finnish
religion traditionally Lutheran
chronology
1918 Estonia declared its independence. March: Soviet forces, who
had tried to regain control from occupying German forces during
World War I, were overthrown by German troops. Nov: Soviet troops
took control after German withdrawal.

1919 Soviet rule overthrown with help of British navy; Estonia declared a democratic republic.

1934 Fascist coup replaced government.

1940 Estonia incorporated into USSR.

1941–44 German occupation during World War II.

1944 USSR regained control.

1980 Beginnings of nationalist dissent.

1988 Adopted own constitution, with power of veto on all centralized Soviet legislation. Popular Front (Rahvarinne) established to campaign for democracy. Estonia's supreme soviet (state assembly) voted to declare the republic 'sovereign' and autonomous in all matters except military and foreign affairs; rejected by USSR as unconstitutional.

1989 Estonian replaced Russian as main language.

1990 Feb: Communist Party monopoly of power abolished; multiparty system established. March: pro-independence candidates secured majority after republic elections; coalition government formed with Popular Front leader Edgar Savisaar as prime minister; Arnold Rüütel became president. May: prewar constitution partially restored.

1991 March: independence plebiscite overwhelmingly approved. Aug: full independence declared after abortive anti-Gorbachev coup; Communist Party outlawed. Sept: independence recognized by Soviet government and Western nations; admitted into United Nations and CSCE (Conference on Security and Cooperation in Europe).

1992 Jan: Savisaar resigned owing to his government's inability to alleviate food and energy shortages; new government formed by Tiit Vahl. June: New constitution approved. Sept: presidential election inconclusive; right-wing Fatherland Group did well in general election. Oct: Meri chosen by parliament to replace Rüütel.

Ethiopia
People's
Democratic
Republic of
(*Hebretesebawit
Ityopia*, formerly
also known as
Abyssinia)

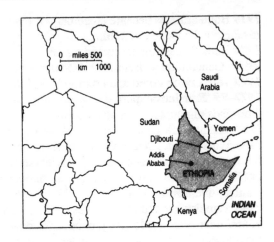

area 1,221,900 sq km/471,653 sq mi
capital Addis Ababa
towns Asmara (capital of Eritrea), Dire Dawa, Massawa, Assab
physical a high plateau with central mountain range divided by Rift
Valley; plains in E; source of Blue Nile River
environment more than 90% of the forests of the Ethiopian highlands
have been destroyed since 1900
head of state and government Meles Zenawi from 1991
political system transition to democratic socialist republic
exports coffee, pulses, oilseeds, hides, skins
currency birr
population (1989 est) 47,709,000 (Oromo 40%, Amhara 25%, Tigré
12%, Sidamo 9%); growth rate 2.5% p.a.
life expectancy 38
languages Amharic (official), Tigrinya, Orominga, Arabic
religions Sunni Muslim 45%, Christian (Ethiopian Orthodox Church,
which has had its own patriarch since 1976) 40%
literacy 35% (1988)
GDP $4.8 bn (1987); $104 per head
chronology
1889 Abyssinia reunited by Menelik II.

1930 Haile Selassie became emperor.

1962 Eritrea annexed by Haile Selassie; resistance movement began.

1974 Haile Selassie deposed and replaced by a military government led by General Teferi Benti. Ethiopia declared a socialist state.

1977 Teferi Benti killed and replaced by Col Mengistu Haile Mariam.

1977–79 'Red Terror' period in which Mengistu's regime killed thousands of people.

1981–85 Ethiopia spent at least $2 billion on arms.

1985 Worst famine in more than a decade; Western aid sent and forcible internal resettlement programmes undertaken.

1987 New constitution adopted, Mengistu Mariam elected president. New famine; food aid hindered by guerrillas.

1988 Mengistu agreed to adjust his economic policies in order to secure International Monetary Fund assistance. Influx of refugees from Sudan.

1989 Coup attempt against Mengistu foiled. Peace talks with Eritrean rebels mediated by former US president Carter reported some progress.

1990 Rebels captured port of Massawa. Mengistu announced new reforms.

1991 Mengistu overthrown; transitional government set up by Ethiopian People's Revolutionary Democratic Front (EPRDF). Eritrean People's Liberation Front (EPLF) secured Eritrea; Eritrea's right to secede recognized; Meles Zenawi elected Ethiopia's new head of state and government. Isaias Afwerki became secretary general of provisional government in Eritrea.

1993 April: overwhelming majority voted in favour of Eritrean independence in referendum.

Fiji
Republic of

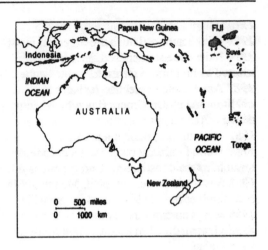

area 18,333 sq km/7,078 sq mi
capital Suva
towns Lautoka, Levuka
physical comprises 844 Melanesian and Polynesian islands and islets
(about 110 inhabited), the largest being Viti Levu (10,429 sq km/
4,028 sq mi) and Vanua Levu (5,550 sq km/2,146 sq mi);
mountainous, volcanic, with tropical rainforest and grasslands
head of state Ratu Sir Peñaia Ganilau from 1987
head of government Col Sitiveni Rabuka from 1992
political system democratic republic
exports sugar, coconut oil, ginger, timber, canned fish, gold; tourism
is important
currency Fiji dollar
population (1989 est) 758,000 (46% Fijian, holding 80% of the land
communally, and 49% Indian, introduced in the 19th century to work
the sugar crop); growth rate 2.1% p.a.
life expectancy men 67, women 71
languages English (official), Fijian, Hindi
religions Hindu 50%, Methodist 44%
literacy men 88%, women 77% (1980 est)
GDP $1.2 bn (1987); $1,604 per head

chronology
1874 Fiji became a British crown colony.
1970 Independence achieved from Britain; Ratu Sir Kamisese Mara
elected as first prime minister.
1987 April: general election brought to power an Indian-dominated
coalition led by Dr Timoci Bavadra. May: military coup by Col
Sitiveni Rabuka removed new government at gunpoint; Governor
General Ratu Sir Penaia Ganilau regained control within weeks.
Sept: second military coup by Rabuka proclaimed Fiji a republic and
suspended the constitution. Oct: Fiji ceased to be a member of the
Commonwealth. Dec: civilian government restored with Rabuka
retaining control of security as minister for home affairs.
1990 New constitution, favouring indigenous Fijians, introduced.
1992 General election produced coalition government; Col Rabuka
named as president.

Finland
Republic of
(*Suomen Tasavalta*)

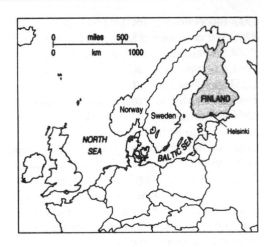

area 338,145 sq km/ 130,608 sq mi
capital Helsinki
towns Tampere, Rovaniemi, Lahti; ports Turku, Oulu
physical most of the country is forest, with low hills and about 60,000
lakes; one-third is within the Arctic Circle; archipelago in S; includes
Åland Islands
head of state Mauno Koivisto from 1982
head of government Esko Aho from 1991
political system democratic republic
exports metal, chemical, and engineering products (icebreakers and oil
rigs), paper, sawn wood, clothing, fine ceramics, glass, furniture
currency markka
population (1989 est) 4,990,000; growth rate 0.5% p.a.
life expectancy men 70, women 78
languages Finnish 93%, Swedish 6% (both official), small Saami- and
Russian-speaking minorities
religions Lutheran 97%, Eastern Orthodox 1.2%
literacy 99%
GDP $77.9 bn (1987); $15,795 per head
chronology
1809 Finland annexed by Russia.

1917 Independence declared from Russia.

1920 Soviet regime acknowledged independence.

1939 Defeated by USSR in Winter War.

1941 Allowed Germany to station troops in Finland to attack USSR; USSR bombed Finland.

1944 Concluded separate armistice with USSR.

1948 Finno-Soviet Pact of Friendship, Cooperation, and Mutual Assistance signed.

1955 Finland joined the United Nations and the Nordic Council.

1956 Urho Kekkonen elected president; re-elected 1962, 1968, 1978.

1973 Trade treaty with European Economic Community signed.

1977 Trade agreement with USSR signed.

1982 Mauno Koivisto elected president; re-elected 1988.

1989 Finland joined Council of Europe.

1991 Big swing to the centre in general election. New coalition government formed.

1992 Formal application for European Community membership.

France
French Republic
(*République
Française*)

area (including island of Corsica) 543,965 sq km/ 209,970 sq mi
capital Paris
towns Lyons, Bordeaux, Toulouse, Strasbourg, Marseille, Nice
physical rivers Seine, Loire, Garonne, Rhône, Rhine; mountain ranges
Alps, Massif Central, Pyrenees, Jura, Vosges, Cévennes
territories Guadeloupe, French Guiana, Martinique, Réunion, St Pierre
and Miquelon, Southern and Antarctic Territories, New Caledonia,
French Polynesia, Wallis and Futuna
head of state François Mitterrand from 1981
head of government Edouard Balladur from 1993
political system liberal democracy
exports fruit (especially apples), wine, cheese, wheat, cars, aircraft,
iron and steel, petroleum products, chemicals, jewellery, silk, lace;
tourism is very important
currency franc
population (1990 est) 56,184,000 (including 4,500,000 immigrants,
chiefly from Portugal, Algeria, Morocco, and Tunisia); growth
rate 0.3% p.a.

life expectancy men 71, women 79
language French; regional languages include Basque, Breton, Catalan,
and the Provençal dialect
religions Roman Catholic 90%, Protestant 2%, Muslim 1%
literacy 99% (1984)
GNP $568 bn (1983); $7,179 per head
recent chronology
1944–46 Provisional government headed by General Charles de
Gaulle; start of Fourth Republic.
1954 Indochina achieved independence.
1956 Morocco and Tunisia achieved independence.
1957 Entry into European Economic Community.
1958 Recall of de Gaulle after Algerian crisis; start of Fifth Republic.
1959 De Gaulle became president.
1962 Algeria achieved independence.
1966 France withdrew from military wing of NATO.
1968 'May events' crisis.
1969 De Gaulle resigned after referendum defeat; Georges Pompidou
became president.
1974 Giscard d'Estaing elected president.
1981 Mitterrand elected Fifth Republic's first socialist president.
1986 'Cohabitation' experiment, with the conservative Jacques Chirac
as prime minister.
1988 Mitterrand re-elected. Moderate socialist Michel Rocard became
prime minister. Matignon Accord on future of New Caledonia
approved by referendum.
1989 Greens gained 11% of vote in elections to European Parliament.
1991 French forces were part of the US-led coalition in the Gulf War.
Edith Cresson became France's first woman prime minister.
Mitterrand's popularity rating fell rapidly.
1992 March: Socialist Party humiliated in regional and local elections;
Greens and National Front polled strongly. April: Cresson replaced by
Pierre Bérégovoy. Sept: referendum narrowly endorsed the
Maastricht Treaty.
1993 March: Socialist Party suffered heavy defeat in National
Assembly elections. Edouard Balladur appointed prime minister.

Gabon
Gabonese Republic
(*République
Gabonaise*)

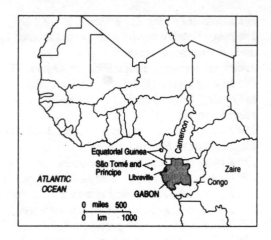

area 267,667 sq km/103,319 sq mi
capital Libreville
towns Port-Gentil, Owendo, Masuku (Franceville)
physical virtually the whole country is tropical rainforest; narrow
coastal plain rising to hilly interior with savanna in E and S; Ogooué
River flows N–W
head of state and government Omar Bongo from 1967
political system emergent democracy
exports petroleum, manganese, uranium, timber
currency CFA franc
population (1988) 1,226,000 including 40 Bantu groups; growth rate
1.6% p.a.
life expectancy men 47, women 51
languages French (official), Bantu
religions Christian 96% (Roman Catholic 65%), small Muslim
minority 1%, animist 3%.
literacy men 70%, women 53% (1985 est)
GDP $3.5 bn (1987); $3,308 per head
chronology
1889 Gabon became part of the French Congo.
1960 Independence from France; Léon M'ba became the first

president.
1967 Attempted coup by rival party foiled with French help. M'ba died; he was succeeded by his protégé Albert-Bernard Bongo.
1968 One-party state established.
1973 Bongo re-elected; converted to Islam, he changed his first name to Omar.
1986 Bongo re-elected.
1989 Coup attempt against Bongo defeated.
1990 Gabonese Democratic Party (PDG) won first multiparty elections since 1964 amidst allegations of ballot-rigging.

Gambia
Republic of The

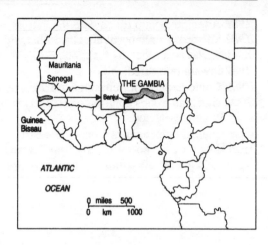

area 10,402 sq km/4,018 sq mi
capital Banjul
towns Serekunda, Bakau, Georgetown
physical banks of the river Gambia flanked by low hills
head of state and government Dawda K Jawara from 1970
political system liberal democracy
exports groundnuts, palm oil, fish
currency dalasi
population (1990 est) 820,000; growth rate 1.9% p.a.
life expectancy 42 (1988 est)
languages English (official), Mandinka, Fula and other native tongues
religions Muslim 90%, with animist and Christian minorities
literacy men 36%, women 15% (1985 est)
GDP $189 million (1987); $236 per head
chronology
1843 The Gambia became a crown colony.
1965 Independence achieved from Britain as a constitutional monarchy within the Commonwealth, with Dawda K Jawara as prime minister.
1970 Declared itself a republic, with Jawara as president.
1972 Jawara re-elected.

1981 Attempted coup foiled with the help of Senegal.
1982 Formed with Senegal the Confederation of Senegambia; Jawara re-elected.
1987 Jawara re-elected.
1989 Confederation of Senegambia dissolved.
1990 Gambian troops contributed to the stabilizing force in Liberia.

Georgia
Republic of

area 69,700 sq km/26,911 sq mi
capital Tbilisi
towns Kutaisi, Rustavi, Batumi, Sukhumi
physical largely mountainous with a variety of landscape from the subtropical Black Sea shores to the ice and snow of the crest line of the Caucasus; chief rivers are Kura and Rioni
interim head of state Eduard Shevardnadze from 1992
head of government Tengiz Sigua from 1992
political system transitional
products tea, citrus and orchard fruits, tung oil, tobacco, vines, silk, hydroelectricity
population (1990) 5,500,000 (Georgian 70%, Armenian 8%, Russian 8%, Azeri 6%, Ossetian 3%, Abkhazian 2%)
language Georgian
religion Georgian Church, independent of the Russian Orthodox Church since 1917
chronology
1918–21 Independent republic.
1921 Uprising quelled by Red Army; Soviet republic established.
1922–36 Linked with Armenia and Azerbaijan as the Transcaucasian Republic.

1936 Became separate republic within USSR.

1981–88 Increasing demands for autonomy, spearheaded from 1988 by the Georgian Popular Front.

1989 March–April: Abkhazians demanded secession from Georgia, provoking inter-ethnic clashes. April: Georgian Communist Party (GCP) leadership purged. July: state of emergency imposed in Abkhazia; inter-ethnic clashes in South Ossetia. Nov: economic and political sovereignty declared.

1990 March: GCP monopoly ended. Oct: nationalist coalition triumphed in supreme soviet elections. Nov: Zviad Gamsakhurdia became president. Dec: GCP seceded from Communist Party of USSR; calls for Georgian independence.

1991 April: declared independence. May: Gamsakhurdia popularly elected president. Aug: GCP outlawed and all relations with USSR severed. Sept: anti-Gamsakhurdia demonstrations; state of emergency declared. Dec: Georgia failed to join new Commonwealth of Independent States (CIS).

1992 Jan: Gamsakhurdia fled to Armenia; Sigua appointed prime minister; Georgia admitted into Conference on Security and Cooperation in Europe (CSCE). Eduard Shevardnadze appointed interim president. July: admitted into United Nations (UN). Aug: fighting started between Georgian troops and Abkhazian separatists in Abkhazia in NW. Oct: Shevardnadze elected chair of new parliament. Clashes in South Ossetia and Abkhazia continued.

1993 Inflation at 1,500%.

Germany
Federal Republic of
(*Bundesrepublik
Deutschland*)

area 357,041 sq km/137,853 sq mi
capital Berlin
towns Cologne, Munich, Essen, Frankfurt-am-Main, Dortmund,
Hamburg
physical flat in N, mountainous in S with Alps; rivers Rhine, Weser,
Elbe flow N, Danube flows SE, Oder, Neisse flow N along Polish
frontier; many lakes, including Müritz
environment acid rain causing *Waldsterben* (tree death) affects more
than half the country's forests; industrial E Germany has the highest
sulphur-dioxide emissions in the world per head of population
head of state Richard von Weizsäcker from 1984
head of government Helmut Kohl from 1982
political system liberal democratic federal republic
exports machine tools (world's leading exporter), cars, commercial
vehicles, electronics, industrial goods, textiles, chemicals, iron, steel,
wine, lignite (world's largest producer), uranium, coal, fertilizers,
plastics
currency Deutschmark

population (1990) 78,420,000 (including nearly 5,000,000 'guest workers', *Gastarbeiter*, of whom 1,600,000 are Turks; the rest are Yugoslav, Italian, Greek, Spanish, and Portuguese); growth rate –0.7% p.a.

life expectancy men 68, women 74

languages German, Sorbian

religions Protestant 42%, Roman Catholic 35%

literacy 99% (1985)

GNP $1,250 bn (1989); $16,200 per head

recent chronology

1945 Germany surrendered; country divided into four occupation zones (US, French, British, Soviet).

1948 Blockade of West Berlin.

1949 Establishment of Federal Republic under the 'Basic Law' Constitution with Konrad Adenauer as chancellor; establishment of the German Democratic Republic as an independent state.

1953 Uprising in East Berlin suppressed by Soviet troops.

1954 Grant of full sovereignty to both West Germany and East Germany.

1957 West Germany was a founder-member of the European Economic Community; recovery of the Saarland from France.

1961 Construction of Berlin Wall.

1963 Retirement of Chancellor Adenauer.

1964 Treaty of Friendship and Mutual Assistance signed between East Germany and USSR.

1969 Willy Brandt became chancellor of West Germany.

1971 Erich Honecker elected Socialist Unity Party (SED) leader in East Germany.

1972 Basic Treaty between West Germany and East Germany; treaty ratified 1973, normalizing relations between the two.

1974 Resignation of Brandt; Helmut Schmidt became chancellor.

1975 East German friendship treaty with USSR renewed for 25 years.

1982 Helmut Kohl became West German chancellor.

1987 Official visit of Honecker to the Federal Republic.

1988 Death of Franz-Josef Strauss, leader of the West German Bavarian Christian Social Union (CSU)

1989 West Germany: rising support for far right in local and European elections, declining support for Kohl. East Germany: mass exodus to West Germany began. Honecker replaced by Egon Krenz. National borders opened in Nov, including Berlin Wall. Reformist Hans Modrow appointed prime minister. Krenz replaced.

1990 March: East German multiparty elections won by a coalition led by the right-wing Christian Democratic Union (CDU). 3 Oct: official reunification of East and West Germany. 2 Dec: first all-German elections since 1932, resulting in a victory for Kohl.

1991 Kohl's popularity declined after tax increase. The CDU lost its Bundesrat majority to the Social Democratic Party (SPD). Racism continued with violent attacks on foreigners.

1992 Neo-Nazi riots against immigrants continued.

1993 Jan: unemployment exceeded 7%; 1% decline in national output predicted for 1993. Honecker allowed to leave for exile in Chile. March: support for CDU slumped in state election in Hesse; Republicans captured 8% of vote.

Ghana
Republic of

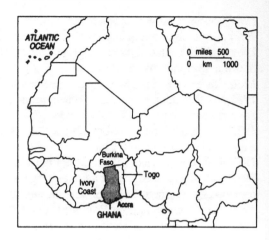

area 238,305 sq km/91,986 sq mi
capital Accra
towns Kumasi, ports Sekondi-Takoradi, Tema
physical mostly tropical lowland plains; bisected by river Volta
environment forested areas have shrunk from 8.2 million sq km/
3.17 million sq mi at the beginning of the 20th century to
1.9 million sq km/730,000 sq mi by 1990
head of state and government Jerry Rawlings from 1981
political system military republic
exports cocoa, coffee, timber, gold, diamonds, manganese, bauxite
currency cedi
population (1990 est) 15,310,000; growth rate 3.2% p.a.
life expectancy men 50, women 54
languages English (official) and African languages
religion animist 38%, Muslim 30%, Christian 24%
literacy men 64%, women 43% (1985 est)
GNP $3.9 bn (1983); $420 per head
chronology
1957 Independence achieved from Britain, within the Commonwealth,
with Kwame Nkrumah as president.
1960 Ghana became a republic.

1964 Ghana became a one-party state.

1966 Nkrumah deposed and replaced by General Joseph Ankrah.

1969 Ankrah replaced by General Akwasi Afrifa, who initiated a return to civilian government.

1970 Edward Akufo-Addo elected president.

1972 Another coup placed Col Acheampong at the head of a military government.

1978 Acheampong deposed in a bloodless coup led by Frederick Akuffo; another coup put Flight-Lt Jerry Rawlings in power.

1979 Return to civilian rule under Hilla Limann.

1981 Rawlings seized power again, citing the incompetence of previous governments. All political parties banned.

1989 Coup attempt against Rawlings foiled.

1992 New multiparty constitution approved. Partial lifting of ban on political parties. Nov: Rawlings won presidency in national elections.

1993 Fourth republic of Ghana formally inaugurated in Rawlings's presence.

Greece
Hellenic Republic
(*Elliniki Dimokratia*)

area 131,957 sq km/50,935 sq mi
capital Athens
towns Larisa, Piraeus, Thessaloníki, Patras, Iráklion
physical mountainous; a large number of islands, notably Crete, Corfu, and Rhodes
environment acid rain and other airborne pollutants are destroying the Classical buildings and ancient monuments of Athens
head of state Constantine Karamanlis from 1990
head of government Constantine Mitsotakis from 1990
political system democratic republic
exports tobacco, fruit, vegetables, olives, olive oil, textiles, aluminium, iron and steel
currency drachma
population (1990 est) 10,066,000; growth rate 0.3% p.a.
life expectancy men 72, women 76
language Greek
religion Greek Orthodox 97%
literacy men 96%, women 89% (1985)
GDP $40.9 bn (1987); $4,093 per head

chronology

1829 Independence achieved from Turkish rule.

1912–13 Balkan Wars; Greece gained much land.

1941–44 German occupation of Greece.

1946 Civil war between royalists and communists; communists defeated.

1949 Monarchy re-established with Paul as king.

1964 King Paul succeeded by his son Constantine.

1967 Army coup removed the king; Col George Papadopoulos became prime minister. Martial law imposed, all political activity banned.

1973 Republic proclaimed, with Papadopoulos as president.

1974 Former premier Constantine Karamanlis recalled from exile to lead government. Martial law and ban on political parties lifted; restoration of the monarchy rejected by a referendum.

1975 New constitution adopted, making Greece a democratic republic.

1980 Karamanlis resigned as prime minister and was elected president.

1981 Greece became full member of European Economic Community. Andreas Papandreou elected Greece's first socialist prime minister.

1983 Five-year military and economic cooperation agreement signed with USA; ten-year economic cooperation agreement signed with USSR.

1985 Papandreou re-elected.

1988 Relations with Turkey improved. Major cabinet reshuffle after mounting criticism of Papandreou.

1989 Papandreou defeated. Tzannis Tzannetakis became prime minister; his all-party government collapsed. Xenophon Zolotas formed new unity government. Papandreou charged with corruption.

1990 New Democracy Party (ND) won half of parliamentary seats in general election but no outright majority; Constantine Mitsotakis became premier; formed new all-party government. Karamanlis re-elected president.

1992 Papandreou acquitted. Greece opposed recognition of independence of the Yugoslav breakaway republic of Macedonia. Decisive parliamentary vote to ratify Maastricht Treaty.

Grenada

area (including the Grenadines) 340 sq km/131 sq mi
capital St George's
towns Grenville, Hillsborough
physical southernmost of the Windward Islands; mountainous
head of state Elizabeth II from 1974, represented by governor general
head of government Nicholas Braithwaite from 1990
political system emergent democracy
exports cocoa, nutmeg, bananas, mace
currency Eastern Caribbean dollar
population (1990 est) 84,000, 84% of black African descent; growth
rate –0.2% p.a.
life expectancy 69
language English (official); some French patois spoken
religion Roman Catholic 60%
literacy 85% (1985)
GDP $139 million (1987); $1,391 per head
chronology
1974 Independence achieved from Britain; Eric Gairy elected
prime minister.

1979 Gairy removed in bloodless coup led by Maurice Bishop; constitution suspended and a People's Revolutionary Government established.

1982 Relations with the USA and Britain deteriorated as ties with Cuba and the USSR strengthened.

1983 After Bishop's attempt to improve relations with the USA, he was overthrown by left-wing opponents. A coup established the Revolutionary Military Council (RMC), and Bishop and three colleagues were executed. The USA invaded Grenada, accompanied by troops from other E Caribbean countries; RMC overthrown, 1974 constitution reinstated.

1984 The newly formed New National Party (NNP) won 14 of the 15 seats in the house of representatives and its leader, Herbert Blaize, became prime minister.

1989 Herbert Blaize lost leadership of NNP, remaining as head of government; he died and was succeeded by Ben Jones.

1990 Nicholas Braithwaite of the National Democratic Congress (NDC) became prime minister.

1991 Integration into Windward Islands confederation proposed.

Guatemala
Republic of
(*República de
Guatemala*)

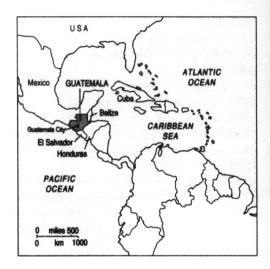

area 108,889 sq km/42,031 sq mi
capital Guatemala City
towns Quezaltenango, Puerto Barrios
physical mountainous; narrow coastal plains; limestone tropical
plateau in N; frequent earthquakes
environment between 1960 and 1980 nearly 57% of the country's
forest was cleared for farming
head of state and government Jorge Serrano Elías from 1991
political system democratic republic
exports coffee, bananas, cotton, sugar, beef
currency quetzal
population (1990 est) 9,340,000 (Mayaquiche Indians 54%, mestizos
(mixed race) 42%); growth rate 2.8% p.a. (87% of under-fives suffer
from malnutrition)
life expectancy men 57, women 61
languages Spanish (official); 40% speak 18 Indian languages
religion Roman Catholic 80%, Protestant 20%
literacy men 63%, women 47% (1985 est)
GDP $7 bn (1987); $834 per head

chronology

1839 Independence achieved from Spain.

1954 Col Carlos Castillo became president in US-backed coup, halting land reform.

1963 Military coup made Col Enrique Peralta president.

1966 Cesar Méndez elected president.

1970 Carlos Araña elected president.

1974 General Kjell Laugerud became president. Widespread political violence precipitated by the discovery of falsified election returns in March.

1978 General Fernando Romeo became president.

1981 Growth of antigovernment guerrilla movement.

1982 General Angel Anibal became president. Army coup installed General Ríos Montt as head of junta and then as president; political violence continued.

1983 Montt removed in coup led by General Mejía Victores, who declared amnesty for the guerrillas.

1985 New constitution adopted; Guatemalan Christian Democratic Party (PDCG) won congressional elections; Vinicio Cerezo elected president.

1989 Coup attempt against Cerezo foiled. Over 100,000 people killed, and 40,000 reported missing since 1980.

1991 Jorge Serrano Elías of the Solidarity Action Movement elected president. Diplomatic relations with Belize established.

Guinea
Republic of
(*République de
Guinée*)

area 245,857 sq km/94,901 sq mi
capital Conakry
towns Labé, Nzérékoré, Kankan
physical flat coastal plain with mountainous interior; sources of rivers
Niger, Gambia, and Senegal; forest in SE
environment large amounts of toxic waste from industrialized
countries have been dumped in Guinea
head of state and government Lansana Conté from 1984
political system military republic
exports coffee, rice, palm kernels, alumina, bauxite, diamonds
currency syli or franc
population (1990 est) 7,269,000 (chief peoples are Fulani, Malinke,
Susu); growth rate 2.3% p.a.
life expectancy men 39, women 42
languages French (official), African languages
religions Muslim 85%, Christian 10%, local 5%
literacy men 40%, women 17% (1985 est)
GNP $1.9 bn (1987); $369 per head
chronology
1958 Full independence achieved from France; Sékou Touré elected
president.

1977 Strong opposition to Touré's rigid Marxist policies forced him to accept return to mixed economy.

1980 Touré returned unopposed for fourth seven-year term.

1984 Touré died. Bloodless coup established a military committee for national recovery, led by Col Lansana Conté.

1985 Attempted coup against Conté while he was out of the country was foiled by loyal troops.

1990 Sent troops to join the multinational force that attempted to stabilize Liberia.

1991 Antigovernment general strike by National Confederation of Guinea Workers (CNTG).

Guinea-Bissau
Republic of
(*República da
Guiné-Bissau*)

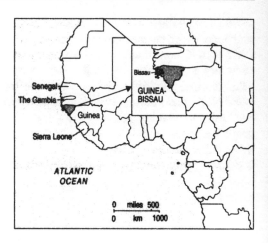

area 36,125 sq km/13,944 sq mi
capital Bissau
towns Mansôa, São Domingos
physical flat coastal plain rising to savanna in E
head of state and government João Bernardo Vieira from 1980
political system socialist pluralist republic
exports rice, coconuts, peanuts, fish, timber
currency peso
population (1989 est) 929,000; growth rate 2.4% p.a.
life expectancy 42; 1990 infant mortality rate was 14.8%
languages Portuguese (official), Crioulo (Cape Verdean dialect of
Portuguese), African languages
religions animism 54%, Muslim 38%, Christian 8%
literacy men 46%, women 17% (1985 est)
GDP $135 million (1987); $146 per head
chronology
1956 African Party for the Independence of Portuguese Guinea and
Cape Verde (PAIGC) formed to secure independence from Portugal.
1973 Two-thirds of the country declared independent, with Luiz
Cabral as president of a state council.
1974 Independence achieved from Portugal.

1980 Cape Verde decided not to join a unified state. Cabral deposed, and João Vieira became chair of a council of revolution.

1981 PAIGC confirmed as the only legal party, with Vieira as its secretary general.

1982 Normal relations with Cape Verde restored.

1984 New constitution adopted, making Vieira head of government as well as head of state.

1989 Vieira re-elected.

1991 Other parties legalized.

1992 Multiparty electoral commission established.

Guyana
Cooperative
Republic of

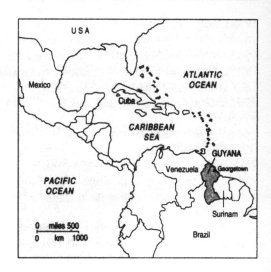

area 214,969 sq km/82,978 sq mi
capital (and port) Georgetown
towns New Amsterdam, Mabaruma
physical coastal plain rises into rolling highlands with savanna in S;
mostly tropical rainforest
head of state Desmond Hoyte from 1985
head of government Cheddi Jagan from 1992
political system democratic republic
exports sugar, rice, rum, timber, diamonds, bauxite, shrimps, molasses
currency Guyanese dollar
population (1989 est) 846,000 (51% descendants of workers
introduced from India to work the sugar plantations after the abolition
of slavery, 30% black, 5% Amerindian); growth rate 2% p.a.
life expectancy men 66, women 71
languages English (official), Hindi, Amerindian
religions Christian 57%, Hindu 33%, Sunni Muslim 9%
literacy men 97%, women 95% (1985 est)
GNP $359 million (1987); $445 per head

chronology

1831 Became British colony under name of British Guiana.

1953 Assembly elections won by left-wing People's Progressive Party (PPP); Britain suspended constitution and installed interim administration, fearing communist takeover.

1961 Internal self-government granted; Cheddi Jagan became prime minister.

1964 People's National Congress (PNC) leader Forbes Burnham led PPP–PNC coalition.

1966 Independence achieved from Britain.

1970 Guyana became a republic within the Commonwealth.

1981 Forbes Burnham became first executive president under new constitution.

1985 Burnham died; succeeded by Desmond Hoyte.

1992 PPP had decisive victory in assembly elections; Jagan returned as prime minister.

Haiti
Republic of
(*République d'Haïti*)

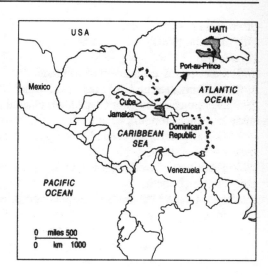

area 27,750 sq km/10,712 sq mi
capital Port-au-Prince
towns Cap-Haïtien, Gonaïves, Les Cayes
physical mainly mountainous and tropical; occupies W third of
Hispaniola Island in Caribbean Sea; seriously deforested
interim head of state Joseph Nerette from 1991
head of government Marc Bazin from 1992
political system transitional
exports coffee, sugar, sisal, cotton, cocoa, bauxite
currency gourde
population (1990 est) 6,409,000; growth rate 1.7% p.a.; one of highest
population densities in the world; about 1.5 million Haitians live
outside Haiti (in USA and Canada); about 400,000 live in virtual
slavery in the Dominican Republic, where they went or were sent to
cut sugar cane
life expectancy men 51, women 54
languages French (official, spoken by literate 10% minority), Creole
(spoken by 90% black majority)
religion Christian 95% (of which 80% Roman Catholic), voodoo 4%

literacy men 40%, women 35% (1985 est)
GDP $2.2 bn (1987); $414 per head
chronology
1804 Independence achieved from France.
1915 Haiti invaded by USA; remained under US control until 1934.
1957 Dr François Duvalier (Papa Doc) elected president.
1964 Duvalier pronounced himself president for life.
1971 Duvalier died, succeeded by his son, Jean-Claude (Baby Doc);
thousands murdered during Duvalier era.
1986 Duvalier deposed; replaced by Lt-Gen Henri Namphy as head of
a governing council.
1988 Feb: Leslie Manigat became president. Namphy staged a military
coup in June, but another coup in Sept led by Brig-Gen Prosper Avril
replaced him with a civilian government under military control.
1989 Coup attempt against Avril foiled; US aid resumed.
1990 Opposition elements expelled; Ertha Pascal-Trouillot acting
president.
1991 Jean-Bertrand Aristide elected president but later overthrown in
military coup led by Brig-Gen Raoul Cedras. Efforts to reinstate
Aristide failed. Joseph Nerette became interim head of state.
1992 Economic sanctions imposed since 1991 were eased by the USA
but increased by the Organization of American States (OAS). Marc
Bazin appointed premier.

Honduras
Republic of
(*República de
Honduras*)

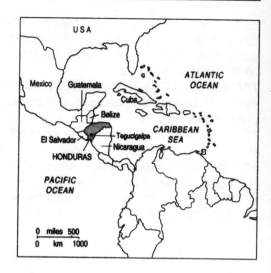

area 112,100 sq km/43,282 sq mi
capital Tegucigalpa
towns San Pedro Sula; ports La Ceiba, Puerto Cortés
physical narrow tropical coastal plain with mountainous interior, Bay
Islands
head of state and government Rafael Leonardo Callejas from 1990
political system democratic republic
exports coffee, bananas, meat, sugar, timber (including mahogany,
rosewood)
currency lempira
population (1989 est) 5,106,000 (mestizo, or mixed, 90%, Indians and
Europeans 10%); growth rate 3.1% p.a.
life expectancy men 58, women 62
languages Spanish (official), English, Indian languages
religion Roman Catholic 97%
literacy men 61%, women 58% (1985 est)
GDP $3.5 bn (1987); $758 per head
chronology
1838 Independence achieved from Spain.

1980 After more than a century of mostly military rule, a civilian government was elected, with Dr Roberto Suazo as president; the commander in chief of the army, General Gustavo Alvarez, retained considerable power.

1983 Close involvement with the USA in providing naval and air bases and allowing Nicaraguan counter-revolutionaries ('Contras') to operate from Honduras.

1984 Alvarez ousted in coup led by junior officers, resulting in policy review towards USA and Nicaragua.

1985 José Azcona elected president after electoral law changed, making Suazo ineligible for presidency.

1989 Government and opposition declared support for Central American peace plan to demobilize Nicaraguan Contras based in Honduras; Contras and their dependents in Honduras in 1989 thought to number about 55,000.

1990 Rafael Callejas, National Party (PN), inaugurated as president.

1992 Border dispute with El Salvador dating from 1861 finally resolved.

Hungary
Republic of
(*Magyar
Köztársaság*)

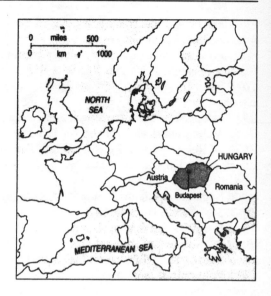

area 93,032 sq km/35,910 sq mi
capital Budapest
towns Miskolc, Debrecen, Szeged, Pécs
physical Great Hungarian Plain covers E half of country; Bakony
Forest, Lake Balaton, and Transdanubian Highlands in the W; rivers
Danube, Tisza, and Raba
environment an estimated 35%–40% of the population live in areas
with officially 'inadmissible' air and water pollution.
head of state Arpád Göncz from 1990
head of government József Antall from 1990
political system emergent democratic republic
exports machinery, vehicles, iron and steel, chemicals, fruit and
vegetables
currency forint
population (1990 est) 10,546,000 (Magyar 92%, Romany 3%,
German 2.5%; Hungarian minority in Romania has caused some
friction between the two countries); growth rate 0.2% p.a.
life expectancy men 67, women 74

language Hungarian (or Magyar), one of the few languages of Europe with non-Indo-European origins
religions Roman Catholic 67%, other Christian denominations 25%
literacy men 99.3%, women 98.5% (1980)
GDP $26.1 bn (1987); $2,455 per head
chronology
1918 Independence achieved from Austro-Hungarian empire.
1919 A communist state formed for 133 days.
1920-44 Regency formed under Admiral Horthy, who joined Hitler's attack on the USSR.
1945 Liberated by USSR.
1946 Republic proclaimed; Stalinist regime imposed.
1949 Soviet-style constitution adopted.
1956 Hungarian national uprising; workers' demonstrations in Budapest; democratization reforms by Imre Nagy overturned by Soviet tanks, János Kádár installed as party leader.
1968 Economic decentralization reforms.
1983 Competition introduced into elections.
1987 VAT and income tax introduced.
1988 Kádár replaced by Károly Grosz. First free trade union recognized; rival political parties legalized.
1989 May: border with Austria opened. July: new four-person collective leadership of the Hungarian Socialist Workers' Party (HSWP). Oct: new 'transitional constitution' adopted, founded on multiparty democracy and new presidentialist executive. HSWP changed name to Hungarian Socialist Party (HSP), with Nyers as new leader. Kádár 'retired'.
1990 HSP reputation damaged by 'Danubegate' bugging scandal. March–April: elections won by right-of-centre coalition, headed by Hungarian Democratic Forum (MDF). May: József Antall, leader of the MDF, appointed premier. Aug: Arpád Göncz elected president.
1991 Jan: devaluation of currency. June: legislation approved to compensate owners of land and property expropriated under communist government. Last Soviet troops departed. Dec: EC association pact signed.
1992 March: EC pact came into effect.

Iceland
Republic of
(*Ly´dveldid Ísland*)

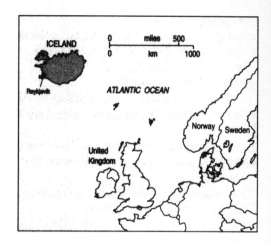

area 103,000 sq km/39,758 sq mi
capital Reykjavík
towns Akureyri, Akranes
physical warmed by the Gulf Stream; glaciers and lava fields cover
75% of the country; active volcanoes (Hekla was once thought the
gateway to Hell), geysers, hot springs, and new islands created
offshore (Surtsey in 1963); subterranean hot water heats 85% of
Iceland's homes
head of state Vigdís Finnbogadóttir from 1980
head of government Davíd Oddsson from 1991
political system democratic republic
exports cod and other fish products, aluminium, diatomite
currency krona
population (1990 est) 251,000; growth rate 0.8% p.a.
life expectancy men 74, women 80
language Icelandic, the most archaic Scandinavian language, in which
some of the finest sagas were written
religion Evangelical Lutheran 95%
literacy 99.9% (1984)
GDP $3.9 bn (1986); $16,200 per head

chronology

1944 Independence achieved from Denmark.

1949 Joined NATO and Council of Europe.

1953 Joined Nordic Council.

1976 'Cod War' with UK.

1979 Iceland announced 320-km/200-mi exclusive fishing zone.

1983 Steingrímur Hermannsson appointed to lead a coalition government.

1985 Iceland declared itself a nuclear-free zone.

1987 New coalition government formed by Thorsteinn Pálsson after general election.

1988 Vigdís Finnbogadóttir re-elected president for a third term; Hermannsson led new coalition.

1991 Davíd Oddsson led new IP–SDP (Independence Party and Social Democratic Party) centre-right coalition, becoming prime minister in the general election.

1992 Iceland defied world ban to resume whaling industry.

India
Republic of
(Hindi *Bharat*)

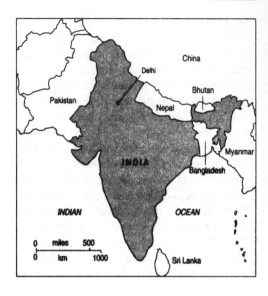

area 3,166,829 sq km/1,222,396 sq mi
capital Delhi
towns Bangalore, Hyderabad, Ahmedabad, Calcutta, Bombay, Madras
physical Himalaya mountains on N border; plains around rivers
Ganges, Indus, Brahmaputra; Andaman and Nicobar Islands,
Lakshadweep (Laccadive Islands)
environment the Narmada Valley Project is the world's largest
hydroelectric/irrigation scheme. In addition to displacing a million
people, the damming of the holy Narmada River will submerge large
areas of forest and farmland and create problems of salinization
head of state Shankar Dayal Sharma from 1992
head of government P V Narasimha Rao from 1991
political system liberal democratic federal republic
exports tea (world's largest producer), coffee, fish, iron and steel,
leather, textiles, clothing, polished diamonds
currency rupee
population (1991 est) 844,000,000 (920 women to every 1,000 men);
growth rate 2.0% p.a.

life expectancy men 56, women 55
languages Hindi, English, and 14 other official languages: Assamese, Bengali, Gujarati, Kannada, Kashmiri, Malayalam, Marathi, Oriya, Punjabi, Sanskrit, Sindhi, Tamil, Telugu, Urdu
religions Hindu 80%, Sunni Muslim 10%, Christian 2.5%, Sikh 2%
literacy men 57%, women 29% (1985 est)
GDP $220.8 bn (1987); $283 per head
chronology
1947 Independence achieved from Britain.
1950 Federal republic proclaimed.
1964 Death of Nehru. Border war with Pakistan over Kashmir.
1966 Indira Gandhi became prime minister.
1971 War with Pakistan leading to creation of Bangladesh.
1975–77 State of emergency proclaimed.
1977–79 Janata Party government in power.
1980 Indira Gandhi returned in landslide victory.
1984 Indira Gandhi assassinated; her son Rajiv Gandhi elected with record majority.
1987 Public revelation of Bofors corruption scandal.
1988 New opposition party, Janata Dal, established by former finance minister V P Singh. Voting age lowered from 21 to 18.
1989 Congress (I) lost majority in general election; Janata Dal minority government formed, with V P Singh prime minister.
1990 Central rule imposed in Jammu and Kashmir. V P Singh resigned; new minority Janata Dal government formed by Chandra Shekhar. Interethnic and religious violence in Punjab and elsewhere.
1991 Central rule imposed in Tamil Nadu. Shekhar resigned; elections called for May. May: Rajiv Gandhi assassinated. June: elections resumed, resulting in a Congress (I) minority government led by P V Narasimha Rao. Separatist violence continued.
1992 Congress (I) won control of state assembly and a majority in parliament in Punjab state elections. Split in Janata Dal opposition resulted in creation of National Front coalition party. Widespread communal violence killed over 1,200 people, mainly Muslims, following destruction of a mosque in Ayodhya by Hindu extremists.
1993 Sectarian violence in Bombay left 500 dead.

Indonesia
Republic of
(*Republik Indonesia*)

area 1,919,443 sq km/740,905 sq mi
capital Jakarta
towns Bandung, Surabaya, Semarang, Tandjungpriok
physical comprises 13,677 tropical islands, of the Greater Sunda group
(including Java and Madura, part of Borneo (Kalimantan), Sumatra,
Sulawesi and Belitung), and the Lesser Sundas/Nusa Tenggara
(including Bali, Lombok, Sumbawa, Sumba, Flores, and Timor), as
well as Malaku/Moluccas and part of New Guinea (Irian Jaya)
environment comparison of primary forest and 30-year-old secondary
forest has shown that logging in Kalimantan has led to a 20% decline
in tree species
head of state and government T N J Suharto from 1967
political system authoritarian nationalist republic
exports coffee, rubber, timber, palm oil, coconuts, tin, tea, tobacco, oil,
liquid natural gas
currency rupiah
population (1989 est) 187,726,000 (including 300 ethnic groups);
growth rate 2% p.a.; Indonesia has the world's largest Muslim
population; Java is one of the world's most densely populated areas

life expectancy men 52, women 55
languages Indonesian (official), closely allied to Malay; Javanese is the most widely spoken local dialect
religions Muslim 88%, Christian 10%, Buddhist and Hindu 2%
literacy men 83%, women 65% (1985 est)
GDP $69.7 bn (1987); $409 per head
chronology
1942 Occupied by Japan; nationalist government established.
1945 Japanese surrender; nationalists declared independence under Achmed Sukarno.
1949 Formal transfer of Dutch sovereignty.
1950 Unitary constitution established.
1963 Western New Guinea (Irian Jaya) ceded by the Netherlands.
1965–66 Attempted communist coup; General T N J Suharto imposed emergency administration, carried out massacre of hundreds of thousands.
1967 Sukarno replaced as president by Suharto.
1975 Guerrillas seeking independence for S Moluccas seized train and Indonesian consulate in the Netherlands, held Western hostages.
1976 Forced annexation of former Portuguese colony of East Timor.
1986 Institution of 'transmigration programme' to settle large numbers of Javanese on sparsely populated outer islands, particularly Irian Jaya.
1988 Partial easing of travel restrictions to East Timor. Suharto re-elected for fifth term.
1989 Foreign debt reaches $50 billion; Western creditors offer aid on condition that concessions are made to foreign companies and that austerity measures are introduced.
1991 Democracy forums launched to promote political dialogue. Massacre in East Timor.
1992 The ruling Golkar party won the assembly elections.
1993 President Suharto re-elected for sixth consecutive five-year term.

Iran
Islamic Republic of
(*Jomhori-e-Islami-e-Irân*; until 1935
Persia)

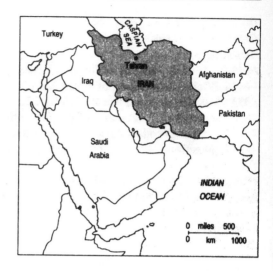

area 1,648,000 sq km/636,128 sq mi
capital Tehran
towns Isfahan, Mashhad, Tabriz, Shiraz, Ahvaz, Abadan
physical plateau surrounded by mountains, including Elburz and
Zagros; Lake Rezayeh; Dasht-Ekavir Desert; occupies islands of Abu
Musa, Greater Tunb and Lesser Tunb in the Gulf
Leader of the Islamic Revolution Seyed Ali Khamenei from 1989
head of government Ali Akbar Hoshemi Rafsanjani from 1989
political system authoritarian Islamic republic
exports carpets, cotton textiles, metalwork, leather goods, oil,
petrochemicals, fruit
currency rial
population (1989 est) 51,005,000 (including minorities in Azerbaijan,
Baluchistan, Khuzestan/Arabistan, and Kurdistan); growth rate
3.2% p.a.
life expectancy men 57, women 57
languages Farsi (official), Kurdish, Turkish, Arabic, English, French
religion Shi'ite Muslim (official) 92%, Sunni Muslim 5%, Zoroastrian
2%, Jewish, Baha'i, and Christian 1%

literacy men 62%, women 39% (1985 est)
GDP $86.4 bn (1987); $1,756 per head
chronology
1946 British, US, and Soviet forces left Iran.
1951 Oilfields nationalized by Prime Minister Muhammad Mossadeq.
1953 Mossadeq deposed and the US-backed shah, Muhammed Reza
Shah Pahlavi, took full control of the government.
1975 The shah introduced single-party system.
1978 Opposition to the shah organized from France by Ayatollah
Khomeini.
1979 Shah left the country; Khomeini returned to create Islamic state.
Revolutionaries seized US hostages at embassy in Tehran; US
economic boycott.
1980 Start of Iran–Iraq War.
1981 US hostages released.
1984 Egyptian peace proposals rejected.
1985 Fighting intensified in Iran–Iraq War.
1988 Cease-fire; talks with Iraq began.
1989 Khomeini called for the death of British writer Salman Rushdie.
June: Khomeini died; Ali Khamenei elected interim Leader of the
Revolution; speaker of Iranian parliament Hoshemi Rafsanjani elected
president. Secret oil deal with Israel revealed.
1990 Generous peace terms with Iraq accepted. Normal relations with
UK restored.
1991 Imprisoned British business executive released. Nearly one
million Kurds arrived in Iran from Iraq, fleeing persecution by Saddam
Hussein after the Gulf War.
1992 Pro-Rafsanjani moderates won assembly elections.

Iraq
Republic of
(*al Jumhouriya al
'Iraqia*)

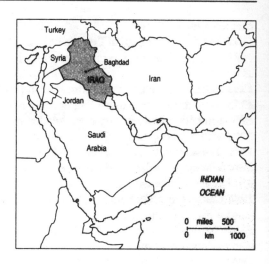

area 434,924 sq km/167,881 sq mi
capital Baghdad
towns Mosul, Basra
physical mountains in N, desert in W; wide valley of rivers Tigris and Euphrates NW–SE
environment a chemical-weapons plant covering an area of 65 sq km/25 sq mi, situated 80 km/50 mi NW of Baghdad, has been described by the UN as the largest toxic waste dump in the world
head of state and government Saddam Hussein al-Tikriti from 1979
political system one-party socialist republic
exports oil (prior to UN sanctions), wool, dates (80% of world supply)
currency Iraqi dinar
population (1989 est) 17,610,000 (Arabs 77%, Kurds 19%, Turks 2%); growth rate 3.6% p.a.
life expectancy men 62, women 63
languages Arabic (official); Kurdish, Assyrian, Armenian
religions Shi'ite Muslim 60%, Sunni Muslim 37%, Christian 3%
literacy men 68%, women 32% (1980 est)
GDP $42.3 bn (1987); $3,000 per head

chronology

1920 Iraq became a British League of Nations protectorate.

1921 Hashemite dynasty established, with Faisal I installed by Britain as king.

1932 Independence achieved from British protectorate status.

1958 Monarchy overthrown; Iraq became a republic.

1963 Joint Ba'athist-military coup headed by Col Salem Aref.

1968 Military coup put Maj-Gen al-Bakr in power.

1979 Al-Bakr replaced by Saddam Hussein.

1980 War between Iraq and Iran broke out.

1985 Fighting intensified.

1988 Cease-fire; talks began with Iran. Iraq used chemical weapons against Kurdish rebels seeking greater autonomy.

1989 Unsuccessful coup against President Hussein; Iraq launched ballistic missile in successful test.

1990 Peace treaty favouring Iran agreed. Aug: Iraq invaded and annexed Kuwait, precipitating another Gulf crisis. US forces massed in Saudi Arabia at request of King Fahd. United Nations resolutions ordered Iraqi withdrawal from Kuwait and imposed total trade ban on Iraq; UN resolution sanctioning force approved. All foreign hostages released.

1991 16 Jan: US-led forces began aerial assault on Iraq; Iraq's infrastructure destroyed by bombing. 23–28 Feb: land–sea–air offensive to free Kuwait successful. Uprisings of Kurds and Shi'ites brutally suppressed by surviving Iraqi troops. Talks between Kurdish leaders and Saddam Hussein about Kurdish autonomy. Allied troops withdrew after establishing 'safe havens' for Kurds in the north, leaving a rapid-reaction force near the Turkish border. Allies threatened to bomb strategic targets in Iraq if full information about nuclear facilities denied to UN.

1992 UN imposed a 'no-fly zone' over S Iraq to protect Shi'ites.

1993 Jan: Iraqi incursions into the 'no-fly zone' prompted US-led alliance aircraft to bomb 'strategic' targets in Iraq. Relations subsequently improved.

Ireland
Republic of
(*Eire*)

area 70,282 sq km/27,146 sq mi
capital Dublin
towns Cork, Dun Laoghaire, Limerick, Waterford
physical central plateau surrounded by hills; rivers Shannon, Liffey,
Boyne
head of state Mary Robinson from 1990
head of government Albert Reynolds from 1992
political system democratic republic
exports livestock, dairy products, Irish whiskey, microelectronic
components and assemblies, mining and engineering products,
chemicals, clothing; tourism is important
currency punt
population (1989 est) 3,734,000; growth rate 0.1% p.a.
life expectancy men 70, women 76
languages Irish Gaelic and English (both official)
religion Roman Catholic 94%
literacy 99% (1984)
GDP $21.9 (1987); $6,184 per head

chronology

1937 Independence achieved from Britain.

1949 Eire left the Commonwealth and became the Republic of Ireland.

1973 Fianna Fáil defeated after 40 years in office; Liam Cosgrave formed a coalition government.

1977 Fianna Fáil returned to power, with Jack Lynch as prime minister.

1979 Lynch resigned, succeeded by Charles Haughey.

1981 Garret FitzGerald formed a coalition.

1983 New Ireland Forum formed, but rejected by the British government.

1985 Anglo-Irish Agreement signed.

1986 Protests by Ulster Unionists against the agreement.

1987 General election won by Charles Haughey.

1988 Relations with UK at low ebb because of disagreement over extradition decisions.

1989 Haughey failed to win majority in general election. Progressive Democrats given cabinet positions in coalition government.

1990 Mary Robinson elected president; John Bruton became Fine Gael leader.

1992 Jan: Haughey resigned after losing parliamentary majority. Feb: Albert Reynolds became Fianna Fáil leader and prime minister. June: National referendum approved ratification of Maastricht Treaty. Nov: Reynolds lost confidence vote; election result inconclusive.

1993 Fianna Fáil–Labour coalition formed.

Israel
State of
(*Medinat Israel*)

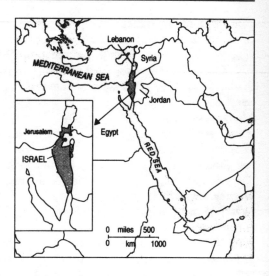

area 20,800 sq km/8,029 sq mi (as at 1949 armistice)
capital Jerusalem (not recognized by the United Nations)
towns ports Tel Aviv/Jaffa, Haifa, Acre, Eilat, Holon, Beersheba
physical coastal plain of Sharon between Haifa and Tel Aviv noted
since ancient times for fertility; central mountains of Galilee, Samariq,
and Judea; Dead Sea, Lake Tiberias, and river Jordan Rift Valley along
the E are below sea level; Negev Desert in the S; Israel occupies Golan
Heights, West Bank, and Gaza
head of state Ezer Weizman from 1993
head of government Yitzhak Rabin from 1992
political system democratic republic
exports citrus and other fruit, fertilizers, diamonds, plastics,
petrochemicals, textiles, electronics (military, medical, scientific,
industrial), precision instruments, aircraft and missiles
currency shekel
population (1989 est) 4,477,000 (including 750,000 Arab Israeli
citizens and over 1 million Arabs in the occupied territories); under the
Law of Return 1950, 'every Jew shall be entitled to come to Israel as
an immigrant'; those from the East and E Europe are Ashkenazim, and

those from Mediterranean Europe (Spain, Portugal, Italy, France, Greece) and Arab N Africa are Sephardim (over 50% of the population is now of Sephardic descent). Between Jan 1990 and April 1991, 250,000 Soviet Jews emigrated to Israel. An Israeli-born Jew is a Sabra. About 500,000 Israeli Jews are resident in the USA. Growth rate 1.8% p.a.

life expectancy men 73, women 76

languages Hebrew and Arabic (official); Yiddish, European and W Asian languages

religions Israel is a secular state, but the predominant faith is Judaism 83%; also Sunni Muslim, Christian, and Druse

literacy Jewish 88%, Arab 70%

GDP $35 bn (1987); $8,011 per head

chronology

1948 Independent State of Israel proclaimed with David Ben-Gurion as prime minister; attacked by Arab nations, Israel won the War of Independence. Many displaced Arabs settled in refugee camps in the Gaza Strip and West Bank.

1952 Col Gamal Nasser of Egypt stepped up blockade of Israeli ports and support of Arab guerrillas in Gaza.

1956 Israel invaded Gaza and Sinai.

1959 Egypt renewed blockade of Israeli trade through Suez Canal.

1963 Ben-Gurion resigned, succeeded by Levi Eshkol.

1964 Palestine Liberation Organization (PLO) founded with the aim of overthrowing the state of Israel.

1967 Israel victorious in the Six-Day War. Gaza, West Bank, E Jerusalem, Sinai, and Golan Heights captured.

1968 Israel Labour Party formed, led by Golda Meir.

1969 Golda Meir became prime minister.

1973 Yom Kippur War: Israel attacked by Egypt and Syria.

1974 Golda Meir succeeded by Yitzhak Rabin.

1975 Suez Canal reopened.

1977 Menachem Begin elected prime minister. Egyptian president addressed the Knesset.

1978 Camp David talks.

1979 Egyptian–Israeli agreement signed. Israel agreed to withdraw from Sinai.

1980 Jerusalem declared capital of Israel.

1981 Golan Heights formally annexed.

1982 Israel pursued PLO fighters into Lebanon.

1983 Peace treaty between Israel and Lebanon signed but not ratified.

1985 Formation of government of national unity with Labour and Likud ministers.

1986 Yitzhak Shamir took over from Peres under power-sharing agreement.

1987 Outbreak of Palestinian uprising (Intifada) in West Bank and Gaza.

1988 Criticism of Israel's handling of Palestinian uprising in occupied territories; PLO acknowledged Israel's right to exist.

1989 New Likud–Labour coalition government formed under Shamir. Limited progress achieved on proposals for negotiations leading to elections in occupied territories.

1990 Coalition collapsed due to differences over peace process. New Shamir right-wing coalition formed.

1991 Shamir gave cautious response to Middle East peace proposals. Some Palestinian prisoners released. Peace talks began in Madrid.

1992 Jan: Shamir lost majority in Knesset when ultra-Orthodox party withdrew from coalition. June: Labour Party, led by Yitzhak Rabin, won elections; coalition formed under Rabin. Aug: US–Israeli loan agreement signed. Dec: Palestinians expelled in face of international criticism.

1993 Jan: UN condemned expulsion of Palestinians. Ban on contacts with PLO formally lifted. Feb: government allowed 100 of 400 expelled Palestinians to return. March: Ezer Weizman elected president; Binyamin 'Bibi' Netanyahu elected leader of Likud party.

Italy
Republic of
(*Repubblica
Italiana*)

area 301,300 sq km/116,332 sq mi
capital Rome
towns Milan, Turin, Naples, Genoa, Palermo, Trieste
physical mountainous (Maritime Alps, Dolomites, Apennines) with
narrow coastal lowlands; rivers Po, Adige, Arno, Tiber, Rubicon;
islands of Sicily, Sardinia, Elba, Capri, Ischia, Lipari, Pantelleria; lakes
Como, Maggiore, Garda
environment Milan has the highest recorded level of sulphur-dioxide
pollution of any city in the world. The Po River, with pollution ten
times higher than officially recommended levels, is estimated to
discharge around 250 tonnes of arsenic into the sea each year
head of state Oscar Luigi Scalfaro from 1992
head of government Carlo Azeglio Ciampi from 1993
political system democratic republic
exports wine (world's largest producer), fruit, vegetables, textiles
(Europe's largest silk producer), clothing, leather goods, motor
vehicles, electrical goods, chemicals, marble (Carrara), sulphur,
mercury, iron, steel
currency lira

population (1990 est) 57,657,000; growth rate 0.1% p.a.
life expectancy men 73, women 80 (1989)
language Italian; German, French, Slovene, and Albanian minorities
religion Roman Catholic 100% (state religion)
literacy 97% (1989)
GDP $748 bn; $13,052 per head (1988)
chronology
1946 Monarchy replaced by a republic.
1948 New constitution adopted.
1954 Trieste returned to Italy.
1976 Communists proposed establishment of broad-based, left–right government, the 'historic compromise'; rejected by Christian Democrats.
1978 Christian Democrat Aldo Moro, architect of the historic compromise, kidnapped and murdered by Red Brigade guerrillas.
1983 Bettino Craxi, a Socialist, became leader of broad coalition government.
1987 Craxi resigned; succeeding coalition fell within months.
1988 Christian Democrats' leader Ciriaco de Mita established a five-party coalition including the Socialists.
1989 De Mita resigned after disagreements within his coalition government; succeeded by Giulio Andreotti. De Mita lost leadership of Christian Democrats; Communists formed 'shadow government'.
1991 Referendum approved electoral reform.
1992 April: ruling coalition lost its majority in general election; President Cossiga resigned, replaced by Oscar Luigi Scalfaro in May. Giuliano Amato, deputy leader of Democratic Party of the Left (PDS), accepted premiership. Sept: lira devalued and Italy withdrew from the Exchange Rate Mechanism.
1993 Feb–March: investigation of corruption network exposed Mafia links with several notable politicians, including Craxi and Andreotti. Craxi resigned Socialist Party leadership; replaced by Giorgio Benvenutu. April: referendum results showed Italian people strongly in favour of majority electoral system. Amato resigned premiership; Carlo Ciampi named as his successor.

Ivory Coast
Republic of
(*République de la
Côte d'Ivoire*)

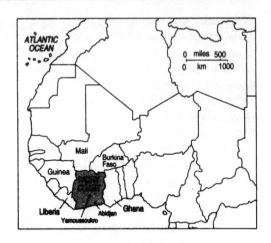

area 322,463 sq km/124,471 sq mi
capital Yamoussoukro
towns Bouaké, Daloa, Man, Abidjan, San-Pédro
physical tropical rainforest (diminishing as exploited) in S; savanna
and low mountains in N
environment an estimated 85% of the country's forest has been
destroyed by humans
head of state and government Félix Houphouët-Boigny from 1960
political system emergent democratic republic
exports coffee, cocoa, timber, petroleum products
currency franc CFA
population (1990 est) 12,070,000; growth rate 3.3% p.a.
life expectancy men 52, women 55 (1989)
languages French (official), over 60 native dialects
religions animist 65%, Muslim 24%, Christian 11%
literacy 35% (1988)
GDP $7.6 bn (1987); $687 per head
chronology
1904 Became part of French West Africa.

1958 Achieved internal self-government.
1960 Independence achieved from France, with Félix Houphouët-Boigny as president of a one-party state.
1985 Houphouët-Boigny re-elected, unopposed.
1986 Name changed officially from Ivory Coast to Côte d'Ivoire.
1990 Houphouët-Boigny and Democratic Party of the Ivory Coast (PDCI) re-elected.

Jamaica

area 10,957 sq km/4,230 sq mi
capital Kingston
towns Montego Bay, Spanish Town, St Andrew
physical mountainous tropical island
head of state Elizabeth II from 1962, represented by governor general
head of government P J Patterson from 1992
political system constitutional monarchy
exports sugar, bananas, bauxite, rum, cocoa, coconuts, liqueurs,
cigars, citrus
currency Jamaican dollar
population (1990 est) 2,513,000 (African 76%, mixed 15%, Chinese,
Caucasian, East Indian); growth rate 2.2% p.a.
life expectancy men 75, women 78 (1989)
languages English, Jamaican creole
religions Protestant 70%, Rastafarian
literacy 82% (1988)
GDP $2.9 bn; $1,187 per head (1989)
chronology
1494 Columbus reached Jamaica.

1509–1655 Occupied by Spanish.

1655 Captured by British.

1944 Internal self-government introduced.

1962 Independence achieved from Britain, with Alexander Bustamante of the Jamaica Labour Party (JLP) as prime minister.

1967 JLP re-elected under Hugh Shearer.

1972 Michael Manley of the People's National Party (PNP) became prime minister.

1980 JLP elected, with Edward Seaga as prime minister.

1983 JLP re-elected, winning all 60 seats.

1988 Island badly damaged by Hurricane Gilbert.

1989 PNP won a decisive victory with Michael Manley returning as prime minister.

1992 Manley resigned, succeeded by P J Patterson.

1993 March: landslide victory for PNP in general election.

Japan
(*Nippon*)

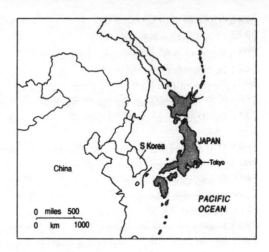

area 377,535 sq km/145,822 sq mi
capital Tokyo
towns Fukuoka, Kyoto, Sapporo,Osaka, Nagoya, Yokohama
physical mountainous, volcanic; comprises over 1,000 islands, the
largest of which are Hokkaido, Honshu, Kyushu, and Shikoku
head of state (figurehead) Emperor Akihito from 1989
head of government Kiichi Miyazawa from 1991
political system liberal democracy
exports televisions, cassette and video recorders, radios, cameras,
computers, robots, other electronic and electrical equipment, motor
vehicles, ships, iron, steel, chemicals, textiles
currency yen
population (1990 est) 123,778,000; growth rate 0.5% p.a.
life expectancy men 76, women 82 (1989)
language Japanese
religions Shinto, Buddhist (often combined), Christian.
literacy 99% (1989)
GDP $2.4 trillion; $19,464 per head (1989)
chronology
1867 End of shogun rule; executive power passed to emperor. Start of
modernization of Japan.

1894–95 War with China; Formosa (Taiwan) and S Manchuria gained.
1902 Formed alliance with Britain.
1904–05 War with Russia; Russia ceded southern half of Sakhalin.
1910 Japan annexed Korea.
1914 Joined Allies in World War I.
1918 Received German Pacific islands as mandates.
1931–32 War with China; renewed 1937.
1941 Japan attacked US fleet at Pearl Harbor 7 Dec.
1945 World War II ended with Japanese surrender. Allied control
commission took power. Formosa and Manchuria returned to China.
1946 Framing of 'peace constitution'. Emperor Hirohito became
figurehead ruler.
1952 Full sovereignty regained.
1958 Joined United Nations.
1968 Bonin and Volcano Islands regained.
1972 Ryukyu Islands regained.
1974 Prime Minister Tanaka resigned over Lockheed bribes scandal.
1982 Yasuhiro Nakasone elected prime minister.
1985 Yen revalued.
1987 Noboru Takeshita chosen to succeed Nakasone.
1988 Recruit scandal cast shadow over government and opposition
parties.
1989 Emperor Hirohito died; succeeded by his son Akihito. Many
cabinet ministers implicated in Recruit scandal and Takeshita resigned;
succeeded by Sosuke Uno. Aug: Uno resigned after sex scandal;
succeeded by Toshiki Kaifu.
1990 Feb: new house of councillors' elections won by the Liberal
Democratic Party (LDP). Public-works budget increased by 50% to
encourage imports.
1991 Japan contributed billions of dollars to the Gulf War and its
aftermath. Kaifu succeeded by Kiichi Miyazawa.
1992 Over 100 politicians implicated in new financial scandal.
Emperor Akihito made first Japanese imperial visit to China. Trade
surpluses reached record levels.
1993 Worst recession of postwar era; trade surpluses, however, again
reached record levels.

Jordan
Hashemite
Kingdom of
(*Al Mamlaka al
Urduniya al
Hashemiyah*)

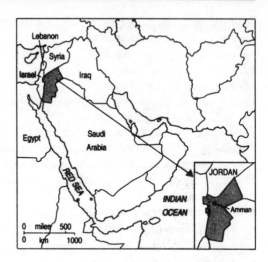

area 89,206 sq km/34,434 sq mi (West Bank 5,879 sq km/2,269 sq mi)
capital Amman
towns Zarqa, Irbid, Aqaba
physical desert plateau in E; rift valley separates E and W banks of the
river Jordan
head of state King Hussein ibn Talai from 1952
head of government Mudar Badran from 1989
political system constitutional monarchy
exports potash, phosphates, citrus, vegetables
currency Jordanian dinar
population (1990 est) 3,065,000 (including Palestinian refugees);
West Bank (1988) 866,000; growth rate 3.6% p.a.
life expectancy men 67, women 71
languages Arabic (official), English
religions Sunni Muslim 92%, Christian 8%
literacy 71% (1988)
GDP $4.3 bn (1987); $1,127 per head (1988)
chronology
1946 Independence achieved from Britain as Transjordan.
1949 New state of Jordan declared.

1950 Jordan annexed West Bank.

1953 Hussein ibn Talai officially became king of Jordan.

1958 Jordan and Iraq formed Arab Federation that ended when the Iraqi monarchy was deposed.

1967 Israel captured and occupied West Bank. Martial law imposed.

1976 Lower house dissolved, political parties banned, elections postponed until further notice.

1982 Hussein tried to mediate in Arab–Israeli conflict.

1984 Women voted for the first time.

1985 Hussein and Yassir Arafat put forward framework for Middle East peace settlement. Secret meeting between Hussein and Israeli prime minister.

1988 Hussein announced decision to cease administering the West Bank as part of Jordan, passing responsibility to Palestine Liberation Organization, and the suspension of parliament.

1989 Prime Minister Zaid al-Rifai resigned; Hussein promised new parliamentary elections following criticism of economic policies. Riots over price increases up to 50% following fall in oil revenues. First parliamentary elections for 22 years; Muslim Brotherhood won 25 of 80 seats but exiled from government; martial law lifted.

1990 Hussein unsuccessfully tried to mediate after Iraq's invasion of Kuwait. Massive refugee problems as thousands fled to Jordan from Kuwait and Iraq.

1991 24 years of martial law ended; ban on political parties lifted.

1992 Political parties allowed to register.

Kazakhstan
Republic of

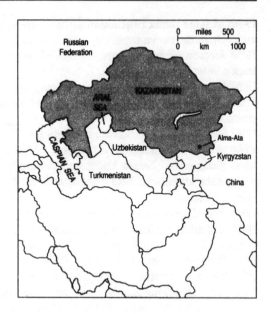

area 2,717,300 sq km/1,049,150 sq mi
capital Alma-Ata
towns Karaganda, Semipalatinsk, Petropavlovsk
physical Caspian and Aral seas, Lake Balkhash; Steppe region
head of state Nursultan Nazarbayev from 1990
head of government Sergey Tereshchenko from 1991
political system emergent democracy
products grain, copper, lead, zinc, manganese, coal, oil
population (1990) 16,700,000 (Kazakh 40%, Russian 38%, German 6%, Ukrainian 5%)
language Russian; Kazakh, related to Turkish
religion Sunni Muslim
chronology
1920 Autonomous republic in USSR.
1936 Joined the USSR and became a full union republic.
1950s Site of Nikita Khrushchev's ambitious 'Virgin Lands' agricultural extension programme.

1960s A large influx of Russian settlers turned the Kazakhs into a minority in their own republic.

1986 Riots in Alma-Alta after Gorbachev ousted local communist leader.

1989 June: Nursultan Nazarbayev became leader of the Kazakh Communist Party (KCP) and instituted economic and cultural reform programmes.

1990 Feb: Nazarbayev became head of state.

1991 March: support pledged for continued union with USSR; Aug: Nazarbayev condemned attempted anti-Gorbachev coup; KCP abolished and replaced by Independent Socialist Party of Kazakhstan (SPK). Dec: joined new Commonwealth of Independent States (CIS); independence recognized by USA.

1992 Jan: admitted into Conference on Security and Cooperation in Europe (CSCE). March: became a member of the United Nations. May: trade agreement with USA.

1993 Jan: new constitution adopted, increasing the authority of the president and making Kazakh the state language.

Kenya
Republic of
(*Jamhuri ya Kenya*)

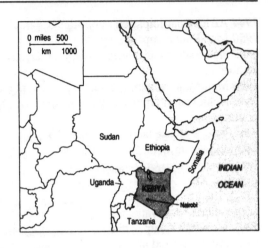

area 582,600 sq km/224,884 sq mi
capital Nairobi
towns Kisumu, Mombasa
physical mountains and highlands in W and centre; coastal plain in S;
arid interior and tropical coast
environment the elephant faces extinction as a result of poaching
head of state and government Daniel arap Moi from 1978
political system authoritarian nationalism
exports coffee, tea, pineapples, petroleum products
currency Kenya shilling
population (1990 est) 25,393,000 (Kikuyu 21%, Luo 13%, Luhya
14%, Kelenjin 11%; Asian, Arab, European); growth rate 4.2% p.a.
life expectancy men 59, women 63 (1989)
languages Kiswahili (official), English; there are many local dialects
religions Protestant 38%, Roman Catholic 28%, indigenous beliefs
26%, Muslim 6%
literacy 50% (1988)
GDP $6.9 bn (1987); $302 per head (1988)
chronology
1895 British East African protectorate established.
1920 Kenya became a British colony.

1944 African participation in politics began.

1950 Mau Mau campaign began.

1953 Nationalist leader Jomo Kenyatta imprisoned by British authorities.

1956 Mau Mau campaign defeated, Kenyatta released.

1963 Achieved internal self-government, with Kenyatta as prime minister.

1964 Independence achieved from Britain as a republic within the Commonwealth, with Kenyatta as president.

1967 East African Community (EAC) formed with Tanzania and Uganda.

1977 Collapse of EAC.

1978 Death of Kenyatta. Succeeded by Daniel arap Moi.

1982 Attempted coup against Moi foiled.

1983 Moi re-elected unopposed.

1984 Over 2,000 people massacred by government forces at Wajir.

1985-86 Thousands of forest villagers evicted and their homes destroyed to make way for cash crops.

1988 Moi re-elected. 150,000 evicted from state-owned forests.

1989 Moi announced release of all known political prisoners. Confiscated ivory burned in attempt to stop elephant poaching.

1990 Despite antigovernment riots, Moi refused multiparty politics.

1991 Increasing demands for political reform; Moi promised multiparty politics.

1992 Constitutional amendment passed. Dec: Moi re-elected in first direct elections despite allegations of fraud.

Kiribati
Republic of

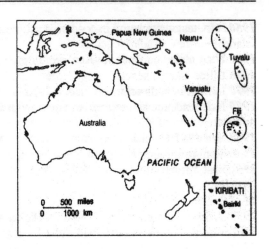

area 717 sq km/277 sq mi
capital (and port) Bairiki (on Tarawa Atoll)
physical comprises 33 Pacific coral islands: the Kiribati (Gilbert), Rawaki (Phoenix), Banaba (Ocean Island), and three of the Line Islands including Kiritimati (Christmas Island)
environment the islands are threatened by the possibility of a rise in sea level caused by global warming. A rise of approximately 30 cm/ 1 ft by the year 2040 will make existing fresh water brackish and undrinkable
head of state and government Teatao Teannaki from 1991
political system liberal democracy
exports copra, fish
currency Australian dollar
population (1990 est) 65,600 (Micronesian); growth rate 1.7% p.a.
languages English (official), Gilbertese
religions Roman Catholic 48%, Protestant 45%
literacy 90% (1985)
GDP $26 million (1987); $430 per head (1988)
chronology
1892 Gilbert and Ellice Islands proclaimed a British protectorate.
1937 Phoenix Islands added to colony.

1950s UK tested nuclear weapons on Kiritimati (formerly Christmas Island).

1962 USA tested nuclear weapons on Kiritimati.

1975 Ellice Islands separated to become Tuvalu.

1977 Gilbert Islands achieved internal self-government.

1979 Independence achieved from Britain, within the Commonwealth, as the Republic of Kiribati, with Ieremia Tabai as president.

1982 and 1983 Tabai re-elected.

1985 Fishing agreement with Soviet state-owned company negotiated, prompting formation of Kiribati's first political party, the opposition Christian Democrats.

1987 Tabai re-elected.

1991 Tabai re-elected but not allowed under constitution to serve further term; Teatao Teannaki elected president.

Korea, North
Democratic
People's Republic of
(*Chosun Minchu-
chui Inmin
Konghwa-guk*)

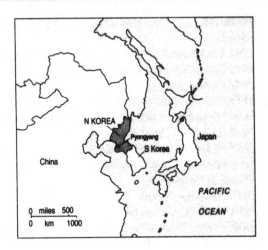

area 120,538 sq km/46,528 sq mi
capital Pyongyang
towns Chongjin, Nampo, Wonsan
physical wide coastal plain in W rising to mountains cut by deep
valleys in interior
environment the building of a hydroelectric dam at Kumgangsan on a
tributary of the Han River has been opposed by South Korea as a
potential flooding threat to central Korea
head of state Kim Il Sung from 1972 (also head of Korean
Workers' Party)
head of government Kang Song San from 1992
political system communism
exports coal, iron, copper, textiles, chemicals
currency won
population (1990 est) 23,059,000; growth rate 2.5% p.a.
life expectancy men 67, women 73 (1989)
language Korean
religions traditionally Buddhist, Confucian, but religious activity
curtailed by the state
literacy 99% (1989)
GNP $20 bn; $3,450 per head (1988)

chronology
1910 Korea formally annexed by Japan.
1945 Russian and US troops entered Korea, forced surrender of Japanese, and divided the country in two. Soviet troops occupied North Korea.
1948 Democratic People's Republic of Korea declared.
1950 North Korea invaded South Korea to unite the nation, beginning the Korean War.
1953 Armistice agreed to end Korean War.
1961 Friendship and mutual assistance treaty signed with China.
1972 New constitution, with executive president, adopted. Talks took place with South Korea about possible reunification.
1980 Reunification talks broke down.
1983 Four South Korean cabinet ministers assassinated in Rangoon, Burma (Myanmar), by North Korean army officers.
1985 Increased relations with the USSR.
1989 Increasing evidence shown of nuclear-weapons development.
1990 Diplomatic contacts with South Korea and Japan suggested the beginning of a thaw in North Korea's relations with the rest of the world.
1991 Became a member of the United Nations. Signed nonaggression agreement with South Korea; agreed to ban nuclear weapons.
1992 Signed Nuclear Safeguards Agreement, allowing international inspection of its nuclear facilities. Also signed a pact with South Korea for mutual inspection of nuclear facilities. Passed legislation making foreign investment in the country attractive. Yon Hyong Muk replaced by Kang Song San.
1993 March: government announced it was pulling out of Nuclear Non-Proliferation Treaty.

Korea, South
Republic of Korea
(*Daehan Minguk*)

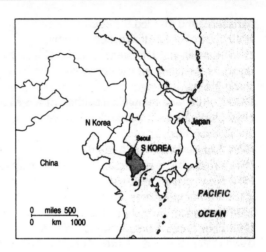

area 98,799 sq km/38,161 sq mi
capital Seoul
towns Taegu, Pusan, Inchon
physical southern end of a mountainous peninsula separating the Sea of Japan from the Yellow Sea
head of state Kim Young Sam from 1992
head of government Hwang In Sung from 1993
political system emergent democracy
exports steel, ships, chemicals, electronics, textiles and clothing, plywood, fish
currency won
population (1990 est) 43,919,000; growth rate 1.4% p.a.
life expectancy men 66, women 73 (1989)
language Korean
religions traditionally Buddhist, Confucian, and Chondokyo; Christian 28%
literacy 92% (1989)
GNP $171bn (1988); $2,180 per head (1986)
chronology
1910 Korea formally annexed by Japan.
1945 Russian and US troops entered Korea, forced surrender of

Japanese, and divided the country in two. US military government took control of South Korea.
1948 Republic proclaimed.
1950-53 War with North Korea.
1960 President Syngman Rhee resigned amid unrest.
1961 Military coup by General Park Chung-Hee. Industrial growth programme.
1979 Assassination of President Park.
1980 Military takeover by General Chun Doo Hwan.
1987 Adoption of more democratic constitution after student unrest. Roh Tae Woo elected president.
1988 Former president Chun, accused of corruption, publicly apologized and agreed to hand over his financial assets to the state. Seoul hosted Summer Olympic Games.
1989 Roh reshuffled cabinet, threatened crackdown on protesters.
1990 Two minor opposition parties united with Democratic Justice Party to form ruling Democratic Liberal Party (DLP). Diplomatic relations established with the USSR.
1991 Violent mass demonstrations against the government. New opposition grouping, the Democratic Party, formed. Prime Minister Ro Jai Bong replaced by Chung Won Shik. Entered United Nations. Nonaggression and nuclear pacts signed with North Korea.
1992 DLP lost absolute majority in March general election; substantial gains made by Democratic Party and newly formed Unification National Party (UNP), led by Chung Ju Wong. Diplomatic relations with China established. Dec: Kim Young Sam, DLP candidate, won the presidential election.
1993 Feb: Kim Young Sam assumed office. Kim Dae Jung and Chung Ju Yung announced their retirement from active politics. Hwang In Sung appointed prime minister.

Kuwait
State of
(*Dowlat al Kuwait*)

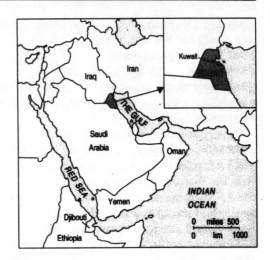

area 17,819 sq km/6,878 sq mi
capital Kuwait
towns Jahra, Ahmadi, Fahaheel
physical hot desert; islands of Failaka, Bubiyan, and Warba at NE corner of Arabian Peninsula
environment during the Gulf War 1990–91, 650 oil wells were set alight and about 300,000 tonnes of oil were released into the waters of the Gulf leading to pollution haze, photochemical smog, acid rain, soil contamination, and water pollution
head of state and government Jabir al-Ahmad al-Jabir al-Sabah from 1977
political system absolute monarchy
exports oil
currency Kuwaiti dinar
population (1990 est) 2,080,000 (Kuwaitis 40%, Palestinians 30%); growth rate 5.5% p.a.
life expectancy men 72, women 76 (1989)
languages Arabic 78%, Kurdish 10%, Farsi 4%
religion Sunni Muslim 45%, Shi'ite minority 30%
literacy 71% (1988)

GNP $19.1 bn; $10,410 per head (1988)
chronology
1914 Britain recognized Kuwait as an independent sovereign state.
1961 Full independence achieved from Britain, with Sheik Abdullah al-Salem al-Sabah as emir.
1965 Sheik Abdullah died; succeeded by his brother, Sheik Sabah.
1977 Sheik Sabah died; succeeded by Crown Prince Jabir.
1983 Shi'ite guerrillas bombed targets in Kuwait; 17 arrested.
1986 National assembly suspended.
1987 Kuwaiti oil tankers reflagged, received US Navy protection; missile attacks by Iran.
1988 Aircraft hijacked by pro-Iranian Shi'ites demanding release of convicted guerrillas; Kuwait refused.
1989 Two of the convicted guerrillas released.
1990 Prodemocracy demonstrations suppressed. Kuwait annexed by Iraq. Emir set up government in exile in Saudi Arabia.
1991 Feb: Kuwait liberated by US-led coalition forces; extensive damage to property and environment. New government omitted any opposition representatives. Trials of alleged Iraqi collaborators criticized.
1992 Oct: reconstituted national assembly elected on restricted franchise, with opposition party winning majority of seats.
1993 Jan: incursions by Iraq into Kuwait again created tension; US-led air strikes restored calm.

Kyrgyzstan
Republic of

area 198,500 sq km/76,641 sq mi
capital Bishkek (formerly Frunze)
towns Osh, Przhevalsk, Kyzyl-Kiya, Tormak
physical mountainous, an extension of the Tian Shan range
head of state Askar Akayev from 1990
head of government Tursunbek Chyngyshev from 1991
political system emergent democracy
products cereals, sugar, cotton, coal, oil, sheep, yaks, horses
population (1990) 4,400,000 (Kyrgyz 52%, Russian 22%, Uzbek
13%, Ukrainian 3%, German 2%)
language Kyrgyz, a Turkic language
religion Sunni Muslim
chronology
1917–1924 Part of an independent Turkestan republic.
1924 Became autonomous republic within USSR.
1936 Became full union republic within USSR.
1990 June: ethnic clashes resulted in state of emergency being
imposed in Bishkek. Nov: Askar Akayev chosen as state president.

1991 March: Kyrgyz voters endorsed maintenance of Union in USSR referendum. Aug: President Akayev condemned anti-Gorbachev attempted coup in Moscow; Kyrgyz Communist Party, which supported the coup, suspended. Oct: Akayev directly elected president. Dec: joined new Commonwealth of Independent States (CIS) and independence recognized by USA.

1992 Jan: admitted into Conference on Security and Cooperation in Europe (CSCE); March: became a member of the United Nations. Dec: Supreme Soviet (parliament) renamed the Uluk Kenesh.

Laos

Lao People's
Democratic
Republic
(*Saathiaranagroat
Prachhathippatay
Prachhachhon Lao*)

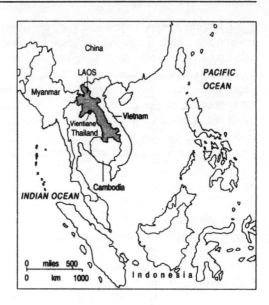

area 236,790 sq km/91,400 sq mi
capital Vientiane
towns Luang Prabang (the former royal capital), Pakse, Savannakhet
physical landlocked state with high mountains in E; Mekong River in
W; jungle covers nearly 60% of land
head of state Nouhak Phoumsavan from 1992
head of government General Khamtay Siphandon from 1991
political system communism, one-party state
exports hydroelectric power from the Mekong is exported to Thailand,
timber, teak, coffee, electricity
currency new kip
population (1990 est) 4,024,000 (Lao 48%, Thai 14%, Khmer 25%,
Chinese 13%); growth rate 2.2% p.a.
life expectancy men 48, women 51 (1989)
languages Lao (official), French
religions Theravāda Buddhist 85%, animist beliefs among mountain
dwellers

literacy 45% (1991)

GNP $500 million (1987); $180 per head (1988)

chronology

1893–1945 Laos was a French protectorate.

1945 Temporarily occupied by Japan.

1946 Retaken by France.

1950 Granted semi-autonomy in French Union.

1954 Independence achieved from France.

1960 Right-wing government seized power.

1962 Coalition government established; civil war continued.

1973 Vientiane cease-fire agreement. Withdrawal of US, Thai, and North Vietnamese forces.

1975 Communist-dominated republic proclaimed with Prince Souphanouvong as head of state.

1986 Phoumi Vongvichit became acting president.

1988 Plans announced to withdraw 40% of Vietnamese forces stationed in the country.

1989 First assembly elections since communist takeover.

1991 Constitution approved. Kaysone Phomvihane elected president. General Khamtay Siphandon named as new premier.

1992 Question of US prisoners of war retained in Laos since the end of Vietnam War unresolved. Nov: Phomvihane died; replaced by Nouhak Phoumsavan. Dec: new national assembly created, replacing supreme people's assembly, and general election held (effectively one-party).

Latvia
Republic of

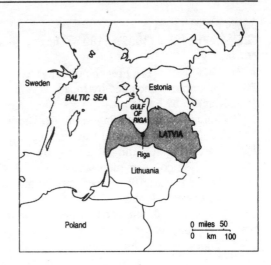

area 63,700 sq km/24,595 sq mi
capital Riga
towns Daugavpils, Liepāja, Jurmala, Jelgava, Ventspils
physical wooded lowland (highest point 312 m/1,024 ft), marshes, lakes; 472 km/293 mi of coastline; mild climate
head of state Anatolijs Gorbunov from 1988
head of government Ivars Godmanis from 1990
political system emergent democratic republic
products electronic and communications equipment, electric railway carriages, motorcycles, consumer durables, timber, paper and woollen goods, meat and dairy products
currency Latvian rouble
population (1990) 2,700,000 (Latvian 52%, Russian 34%, Byelorussian 5%, Ukrainian 3%)
language Latvian
religions mostly Lutheran Protestant, with a Roman Catholic minority
chronology
1917 Soviets and Germans contested for control of Latvia.
1918 Soviet forces overthrown by Germany. Latvia declared independence. Soviet rule restored after German withdrawal.

1919 Soviet rule overthrown by British naval and German forces May–Dec; democracy established.

1934 Coup replaced established government.

1939 German-Soviet secret agreement placed Latvia under Russian influence.

1940 Incorporated into USSR as constituent republic.

1941–44 Occupied by Germany.

1944 USSR regained control.

1980 Nationalist dissent began to grow.

1988 Latvian Popular Front established to campaign for independence. Prewar flag readopted; official status given to Latvian language.

1989 Popular Front swept local elections.

1990 Jan: Communist Party's monopoly of power abolished. March–April: Popular Front secured majority in elections. April: Latvian Communist Party split into pro-independence and pro-Moscow wings. May: unilateral declaration of independence from USSR, subject to transitional period for negotiation.

1991 Jan: Soviet troops briefly seized key installations in Riga. March: overwhelming vote for independence in referendum. Aug: full independence declared at time of anti-Gorbachev coup; Communist Party outlawed. Sept: independence recognized by Soviet government and Western nations; United Nations (UN) membership granted; admitted into Conference on Security and Cooperation in Europe (CSCE).

1992 US reopened its embassy in Latvia. Russia began pullout of ex-Soviet troops, to be completed 1994. July: curbing of rights of non-citizens in Latvia prompted Russia to request minority protection by UN.

Lebanon

Republic of
(*al-Jumhouria al-Lubnaniya*)

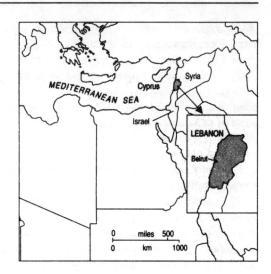

area 10,452 sq km/4,034 sq mi
capital Beirut
towns Tripoli, Tyre, Sidon
physical narrow coastal plain; Bekka valley N–S between Lebanon and Anti-Lebanon mountain ranges
head of state Elias Hrawi from 1989
head of government Rafik al-Hariri from 1992
political system emergent democratic republic
exports citrus and other fruit, vegetables; industrial products to Arab neighbours
currency Lebanese pound
population (1990 est) 3,340,000 (Lebanese 82%, Palestinian 9%, Armenian 5%); growth rate –0.1% p.a.
life expectancy men 65, women 70 (1989)
languages Arabic, French (both official), Armenian, English
religions Muslim 57% (Shiite 33%, Sunni 24%), Christian (Maronite and Orthodox) 40%, Druse 3%
literacy 75% (1989)
GNP $1.8 bn; $690 per head (1986)

chronology

1920–41 Administered under French mandate.

1944 Independence achieved.

1948–49 Lebanon joined first Arab war against Israel. Palestinian refugees settled in the south.

1964 Palestine Liberation Organization (PLO) founded in Beirut.

1971 PLO expelled from Jordan; established headquarters in Lebanon.

1975 Outbreak of civil war between Christians and Muslims.

1976 Cease-fire agreed; Syrian-dominated Arab deterrent force formed to keep the peace but considered by Christians as a occupying force.

1978 Israel invaded S Lebanon in search of PLO fighters. International peacekeeping force established. Fighting broke out again.

1979 Part of S Lebanon declared an 'independent free Lebanon'.

1982 Bachir Gemayel became president but was assassinated before he could assume office; succeeded by his brother Amin Gemayel. Israel again invaded Lebanon. Palestinians withdrew from Beirut under supervision of international peacekeeping force.

1983 Agreement reached for the withdrawal of Syrian and Israeli troops but abrogated under Syrian pressure.

1984 Most of international peacekeeping force withdrawn. Muslim militia took control of W Beirut.

1985 Lebanon in chaos; many foreigners taken hostage.

1987 Syrian troops sent into Beirut.

1989 Christian leader General Michel Aoun declared 'war of liberation' against Syrian occupation; Saudi Arabia and Arab League sponsored talks that resulted in new constitution recognizing Muslim majority; René Muhawad named president, assassinated after 17 days in office; Elias Hrawi named successor; Aoun occupied presidential palace, rejected constitution.

1990 Release of Western hostages began. Aoun surrendered and legitimate government restored, with Umar Karami as prime minister.

1991 Government extended control to the whole country. Treaty of cooperation with Syria signed. More Western hostages released.

1992 Karami succeeded by Rashid al-Solh. Remaining Western hostages released. General election boycotted by many Christians; pro-Syrian administration reelected; Rafik al-Hariri prime minister.

Lesotho
Kingdom of

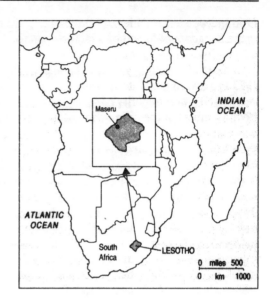

area 30,355 sq km/11,717 sq mi
capital Maseru
towns Teyateyaneng, Mafeteng, Roma, Quthing
physical mountainous with plateaus, forming part of South Africa's
chief watershed
political system emergent democracy
head of state King Letsie III from 1990
head of government Ntsu Mokhehle from 1993
exports wool, mohair, diamonds, cattle, wheat, vegetables
currency maluti
population (1990 est) 1,757,000; growth rate 2.7% p.a.
life expectancy men 59, women 62 (1989)
languages Sesotho, English (official), Zulu, Xhosa
religions Protestant 42%, Roman Catholic 38%
literacy 59% (1988)
GNP $408 million; $410 per head (1988)
chronology
1868 Basutoland became a British protectorate.

1966 Independence achieved from Britain, within the Commonwealth, as the Kingdom of Lesotho, with Moshoeshoe II as king and Chief Leabua Jonathan as prime minister.

1970 State of emergency declared and constitution suspended.

1973 Progovernment interim assembly established; Basotha National Party (BNP) won majority of seats.

1975 Members of the ruling party attacked by guerrillas backed by South Africa.

1985 Elections cancelled because no candidates opposed BNP.

1986 South Africa imposed border blockade, forcing deportation of 60 African National Congress members. General Lekhanya ousted Chief Jonathan in coup. National assembly abolished. Highlands Water Project agreement signed with South Africa.

1990 Moshoeshoe II dethroned by military council; replaced by his son Mohato as King Letsie III.

1991 Lekhanya ousted in military coup led by Col Elias Tutsoane Ramaema. Political parties permitted to operate.

1992 Ex-king Moshoeshoe returned from exile.

1993 Free elections ended military rule; Ntsu Mokhehle of Basutoland Congress Party (BCP) became prime minister.

Liberia
Republic of

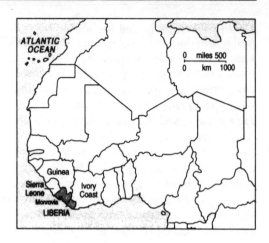

area 111,370 sq km/42,989 sq mi
capital Monrovia
towns Buchanan, Greenville
physical forested highlands; swampy tropical coast where six rivers enter the sea
head of state and government Amos Sawyer from 1990
political system emergent democratic republic
exports iron ore, rubber (Africa's largest producer), timber, diamonds, coffee, cocoa, palm oil
currency Liberian dollar
population (1990 est) 2,644,000 (95% indigenous); growth rate 3% p.a.
life expectancy men 53, women 56 (1989)
languages English (official), over 20 Niger-Congo languages
religions animist 65%, Muslim 20%, Christian 15%
literacy men 47%, women 23% (1985 est)
GNP $973 million; $410 per head (1987)
chronology
1847 Founded as an independent republic.
1944 William Tubman elected president.
1971 Tubman died; succeeded by William Tolbert.

1980 Tolbert assassinated in coup led by Samuel Doe, who suspended
the constitution and ruled through a People's Redemption Council.
1984 New constitution approved. National Democratic Party of
Liberia (NDPL) founded by Doe.
1985 NDPL won decisive victory in general election. Unsuccessful
coup against Doe.
1990 Rebels under former government minister Charles Taylor
controlled nearly entire country by July. Doe killed during a bloody
civil war between rival rebel factions. Amos Sawyer became interim
head of government.
1991 Amos Sawyer re-elected president. Rebel leader Charles Taylor
agreed to work with Sawyer. Peace agreement failed but later revived;
UN peacekeeping force drafted into republic.
1992 Monrovia under siege by Taylor's rebel forces.

Libya
Great Socialist
People's Libyan
Arab Jamahiriya
(*al-Jamahiriya
al-Arabiya al-Libya
al-Shabiya
al-Ishtirakiya
al-Uzma*)

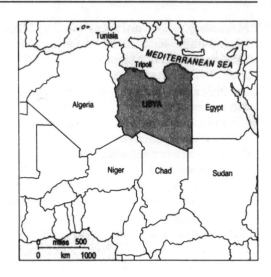

area 1,759,540 sq km/679,182 sq mi
capital Tripoli
towns ports Benghazi, Misurata, Tobruk
physical flat to undulating plains with plateaus and depressions stretch
S from the Mediterranean coast to an extremely dry desert interior
environment plan to pump water from below the Sahara to the coast
risks rapid exhaustion of nonrenewable supply (Great Manmade River
Project)
head of state and government Moamer al-Khaddhafi from 1969
political system one-party socialist state
exports oil, natural gas
currency Libyan dinar
population (1990 est) 4,280,000 (including 500,000 foreign workers);
growth rate 3.1% p.a.
life expectancy men 64, women 69 (1989)
language Arabic
religion Sunni Muslim 97%
literacy 60% (1989)
GNP $20 bn; $5,410 per head (1988)

chronology
1911 Conquered by Italy.
1934 Colony named Libya.
1942 Divided into three provinces: Fezzan (under French control);
Cyrenaica, Tripolitania (under British control).
1951 Achieved independence as the United Kingdom of Libya, under
King Idris.
1969 King deposed in a coup led by Col Moamer al-Khaddhafi.
Revolution Command Council set up and the Arab Socialist Union
(ASU) proclaimed the only legal party.
1972 Proposed federation of Libya, Syria, and Egypt abandoned.
1980 Proposed merger with Syria abandoned. Libyan troops began
fighting in Chad.
1981 Proposed merger with Chad abandoned.
1986 US bombing of Khaddhafi's headquarters, following allegations
of his complicity in terrorist activities.
1988 Diplomatic relations with Chad restored.
1989 USA accused Libya of building a chemical-weapons factory and
shot down two Libyan planes; reconciliation with Egypt.
1992 Khaddhafi under international pressure to extradite suspected
Lockerbie and UTA (Union de Transports Aerians) bombers for trial
outside Libya; sanctions imposed.

Liechtenstein
Principality of
(*Fürstentum
Liechtenstein*)

area 160 sq km/62 sq mi
capital Vaduz
towns Balzers, Schaan, Ruggell
physical landlocked Alpine; includes part of Rhine Valley in W
head of state Prince Hans Adam II from 1989
head of government Hans Brunhart from 1978
political system constitutional monarchy
exports microchips, dental products, small machinery, processed
foods, postage stamps
currency Swiss franc
population (1990 est) 30,000 (33% foreign); growth rate 1.4% p.a.
life expectancy men 78, women 83 (1989)
language German (official); an Alemannic dialect is also spoken
religions Roman Catholic 87%, Protestant 8%
literacy 100% (1989)
GNP $450 million (1986)
chronology
1342 Became a sovereign state.

1434 Present boundaries established.

1719 Former counties of Schellenberg and Vaduz constituted as the Principality of Liechtenstein.

1921 Adopted Swiss currency.

1923 United with Switzerland in a customs union.

1938 Prince Franz Josef II came to power.

1984 Prince Franz Joseph II handed over power to Crown Prince Hans Adam. Vote extended to women in national elections.

1989 Prince Franz Joseph II died; Hans Adam II succeeded him.

1990 Became a member of the United Nations (UN).

1991 Became seventh member of European Free Trade Association.

Lithuania
Republic of

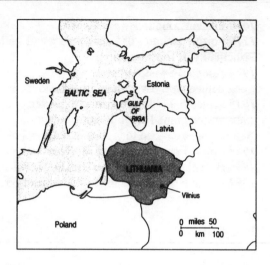

area 65,200 sq km/25,174 sq mi
capital Vilnius
towns Kaunas, Klaipeda, Siauliai, Panevezys
physical central lowlands with gentle hills in W and higher terrain in
SE; 25% forested; some 3,000 small lakes, marshes, and complex
sandy coastline
head of state Algirdas Brazauskas from 1993
head of government Bronislovas Lubys from 1992
political system emergent democracy
products heavy engineering, electrical goods, shipbuilding, cement,
food processing, bacon, dairy products, cereals, potatoes
currency Lithuanian rouble
population (1990) 3,700,000 (Lithuanian 80%, Russian 9%, Polish
7%, Byelorussian 2%)
language Lithuanian
religion predominantly Roman Catholic
chronology
1918 Independence declared following withdrawal of German
occupying troops at end of World War I; USSR attempted to regain
power.

1919 Soviet forces overthrown by Germans, Poles, and nationalist Lithuanians; democratic republic established.

1920–39 Province and city of Vilnius occupied by Poles.

1926 Coup overthrew established government; Antanas Smetona became president.

1939 Secret German-Soviet agreement brought most of Lithuania under Soviet influence.

1940 Incorporated into USSR as constituent republic.

1941 Lithuania revolted against USSR and established own government. During World War II Germany again occupied the country.

1944 USSR resumed rule.

1944–52 Lithuanian guerrillas fought USSR.

1972 Demonstrations against Soviet government.

1980 Growth in nationalist dissent, influenced by Polish example.

1988 Popular front formed, the Sajudis, to campaign for increased autonomy.

1989 Lithuanian declared the state language; flag of independent interwar republic readopted. Communist Party (CP) split into pro-Moscow and nationalist wings. Communist local monopoly of power abolished.

1990 Feb: nationalist Sajudis won elections. March: Vytautas Landsbergis became president; unilateral declaration of independence resulted in temporary Soviet blockade.

1991 Jan: Albertas Shiminas became prime minister. Soviet paratroopers briefly occupied key buildings in Vilnius. Sept: independence recognized by Soviet government and Western nations; Gediminas Vagnorius elected prime minister; CP outlawed; admitted into United Nations and Conference on Security and Cooperation in Europe (CSCE).

1992 July: Aleksandras Abisala became prime minister. Nov: Democratic Labour Party, led by Algirdas Brazauskas, won majority vote. Dec: Bronislovas Lubys appointed prime minister.

1993 Brazauskas elected president.

Luxembourg
Grand Duchy of
(*Grand-Duché de
Luxembourg*)

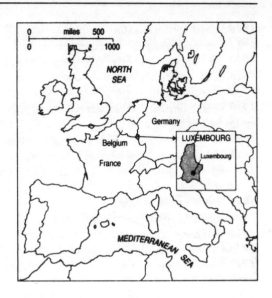

area 2,586 sq km/998 sq mi
capital Luxembourg
towns Esch-sur-Alzette, Dudelange
physical on the river Moselle; part of the Ardennes (Oesling)
forest in N
head of state Grand Duke Jean from 1964
head of government Jacques Santer from 1984
political system liberal democracy
exports pharmaceuticals, synthetic textiles, steel
currency Luxembourg franc
population (1990 est) 369,000; growth rate 0% p.a.
life expectancy men 71, women 78 (1989)
languages French (official), local Letzeburgesch, German
religion Roman Catholic 97%
literacy 100% (1989)
GNP $4.9 bn; $13,380 per head (1988)
chronology
1354 Became a duchy.

1482 Under Habsburg control.

1797 Ceded, with Belgium, to France.

1815 Treaty of Vienna created Luxembourg a grand duchy, ruled by the king of the Netherlands.

1830 With Belgium, revolted against Dutch rule.

1890 Link with Netherlands ended with accession of Grand Duke Adolphe of Nassau-Weilburg.

1948 With Belgium and the Netherlands, formed the Benelux customs union.

1960 Benelux became fully effective economic union.

1961 Prince Jean became acting head of state on behalf of his mother, Grand Duchess Charlotte.

1964 Grand Duchess Charlotte abdicated; Prince Jean became grand duke.

1974 Dominance of Christian Social Party challenged by Socialists.

1979 Christian Social Party regained pre-eminence.

1991 Pact agreeing European free-trade area signed in Luxembourg.

1992 Voted in favour of ratification of Maastricht Treaty on European union.

Madagascar
Democratic Republic of
(*Repoblika
Demokratika n'i
Madagaskar*)

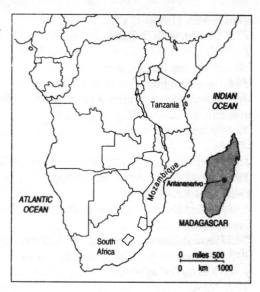

area 587,041 sq km/226,598 sq mi
capital Antananarivo
towns Toamasina, Antseranana, Fianarantsoa, Toliary
physical temperate central highlands; humid valleys and tropical coastal plains; arid in S
environment according to 1990 UN figures, 93% of the forest area has been destroyed and about 100,000 species have been made extinct
head of state Albert Zafy from 1993
head of government Guy Razanamasy from 1991
political system emergent democratic republic
exports coffee, cloves, vanilla, sugar, chromite, shrimps
currency Malagasy franc
population (1990 est) 11,802,000, mostly of Malayo-Indonesian origin; growth rate 3.2% p.a.
life expectancy men 50, women 53 (1989)
languages Malagasy (official), French, English
religions animist 50%, Christian 40%, Muslim 10%

literacy 53% (1988)
GNP $2.1 bn (1987); $280 per head (1988)
chronology
1885 Became a French protectorate.
1896 Became a French colony.
1960 Independence achieved from France, with Philibert Tsiranana as president.
1972 Army took control of the government.
1975 Martial law imposed under a national military directorate. New Marxist constitution proclaimed the Democratic Republic of Madagascar, with Didier Ratsiraka as president.
1976 Front-Line Revolutionary Organization (AREMA) formed.
1977 National Front for the Defence of the Malagasy Socialist Revolution (FNDR) became the sole legal political organization.
1980 Ratsiraka abandoned Marxist experiment.
1983 Ratsiraka re-elected, despite strong opposition from radical socialist National Movement for the Independence of Madagascar (MONIMA) under Monja Jaona.
1989 Ratsiraka re-elected for third term after restricting opposition parties.
1990 Political opposition legalized; 36 new parties created.
1991 Antigovernment demonstrations; opposition to Ratsiraka led to general strike. Nov: Ratsiraka formed new unity government.
1992 Constitutional reform approved. Oct: first multiparty elections won by Democrat coalition.
1993 Albert Zafy, leader of coalition, elected president.

Malawi
Republic of
(*Malaŵi*)

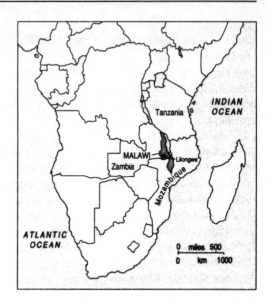

area 118,000 sq km/45,560 sq mi
capital Lilongwe
towns Blantyre, Mzuzu, Zomba
physical landlocked narrow plateau with rolling plains; mountainous
W of Lake Malawi
head of state and government Hastings Kamusu Banda from 1966 for
life
political system one-party republic
exports tea, tobacco, cotton, peanuts, sugar
currency kwacha
population (1990 est) 9,080,000 (nearly 1 million refugees from
Mozambique); growth rate 3.3% p.a.
life expectancy men 46, women 50 (1989)
languages English, Chichewa (both official)
religions Christian 75%, Muslim 20%
literacy 25% (1989)
GNP $1.2 bn (1987); $160 per head (1988)

chronology

1891 Became the British protectorate Nyasaland.

1964 Independence achieved from Britain, within the Commonwealth, as Malawi.

1966 Became a one-party republic, with Hastings Banda as president.

1971 Banda was made president for life.

1977 Banda released some political detainees and allowed greater freedom of the press.

1986–89 Influx of nearly a million refugees from Mozambique.

1992 Calls for multiparty politics. Countrywide industrial riots caused many fatalities. Western aid suspended over human-rights violations. Referendum on constitutional reform promised.

1993 Commission appointed to supervise preparations for the referendum.

Malaysia

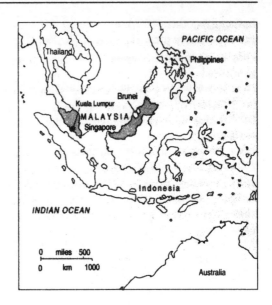

area 329,759 sq km/127,287 sq mi
capital Kuala Lumpur
towns Johor Baharu, Ipoh, Georgetown (Penang), Kuching in
Sarawak, Kota Kinabalu in Sabah
physical comprises Peninsular Malaysia (the nine Malay states—
Johore, Kedah, Kelantan, Negri Sembilan, Pahang, Perak, Perlis,
Selangor, Trengganu—plus Malacca and Penang); and E Malaysia
(Sabah and Sarawak); 75% tropical jungle; central mountain range;
swamps in E
head of state Rajah Azlan Muhibuddin Shah (sultan of Perak)
from 1989
head of government Mahathir bin Mohamad from 1981
political system liberal democracy
exports pineapples, palm oil, rubber, timber, petroleum (Sarawak),
bauxite
currency ringgit
population (1990 est) 17,053,000 (Malaysian 47%, Chinese 32%,

Indian 8%, others 13%); growth rate 2% p.a.
life expectancy men 65, women 70 (1989)
languages Malay (official), English, Chinese, Indian, and local languages
religions Muslim (official), Buddhist, Hindu, local beliefs
literacy 80% (1989)
GNP $34.3 bn; $1,870 per head (1988)
chronology
1786 Britain established control.
1826 Became a British colony.
1963 Federation of Malaysia formed, including Malaya, Singapore, Sabah (N Borneo), and Sarawak (NW Borneo).
1965 Secession of Singapore from federation.
1969 Anti-Chinese riots in Kuala Lumpur.
1971 Launch of *bumiputra*, ethnic-Malay-oriented economic policy.
1981 Election of Dr Mahathir bin Mohamad as prime minister.
1982 Mahathir bin Mohamad re-elected.
1986 Mahathir bin Mohamad re-elected.
1987 Arrest of over 100 opposition activists, including Democratic Action Party (DAP) leader, as Malay-Chinese relations deteriorated.
1988 Split in ruling New United Malays' National Organization (UMNO) party over Mahathir's leadership style; new UMNO formed.
1989 Semangat '46 set up by former members of UMNO including ex-premier Tunku Abdul Rahman.
1990 Mahathir bin Mohamad re-elected.
1991 New economic growth programme launched.

Maldives
Republic of
(*Divehi Jumhuriya*)

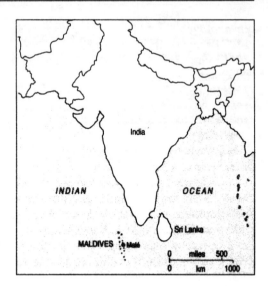

area 298 sq km/115 sq mi
capital Malé
towns Seenu
physical comprises 1,196 coral islands, grouped into 12 clusters of
atolls, largely flat, none bigger than 13 sq km/5 sq mi, average
elevation 1.8 m/6 ft; 203 are inhabited
environment the threat of rising sea level has been heightened by the
frequency of flooding in recent years
head of state and government Maumoon Abdul Gayoom from 1978
political system authoritarian nationalism
exports coconuts, copra, bonito (fish related to tuna), garments
currency Rufiya
population (1990 est) 219,000; growth rate 3.7% p.a.
life expectancy men 60, women 63 (1989)
languages Divehi (Sinhalese dialect), English
religion Sunni Muslim
literacy 36% (1989)
GNP $69 million (1987); $410 per head (1988)

chronology

1887 Became a British protectorate.

1953 Long a sultanate, the Maldive Islands became a republic within the Commonwealth.

1954 Sultan restored.

1965 Achieved full independence outside the Commonwealth.

1968 Sultan deposed; republic reinstated with Ibrahim Nasir as president.

1978 Nasir retired; replaced by Maumoon Abdul Gayoom.

1982 Rejoined the Commonwealth.

1983 Gayoom re-elected.

1985 Became a founder member of South Asian Association for Regional Cooperation (SAARC).

1988 Gayoom re-elected. Coup attempt by mercenaries, thought to have the backing of former president Nasir, was foiled by Indian paratroops.

Mali
Republic of
(*République du Mali*)

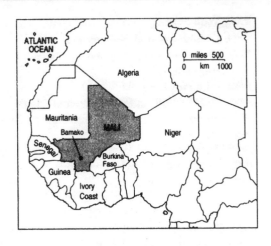

area 1,240,142 sq km/478,695 sq mi
capital Bamako
towns Mopti, Kayes, Ségou, Timbuktu
physical landlocked state with river Niger and savanna in S; part of the Sahara in N; hills in NE; Senegal River and its branches irrigate the SW
environment a rising population coupled with recent droughts has affected marginal agriculture. Once in surplus, Mali has had to import grain every year since 1965
head of state and government Alpha Oumar Konare from 1992
political system emergent democratic republic
exports cotton, peanuts, livestock, fish
currency franc CFA
population (1990 est) 9,182,000; growth rate 2.9% p.a.
life expectancy men 44, women 47 (1989)
languages French (official), Bambara
religion Sunni Muslim 90%, animist 9%, Christian 1%
literacy 10% (1989)
GNP $1.6 bn (1987); $230 per head (1988)
chronology
1895 Came under French rule.

1959 With Senegal, formed the Federation of Mali.

1960 Became the independent Republic of Mali, with Modibo Keita as president.

1968 Keita replaced in an army coup by Moussa Traoré.

1974 New constitution made Mali a one-party state.

1976 New national party, the Malian People's Democratic Union, announced.

1983 Agreement between Mali and Guinea for eventual political and economic integration signed.

1985 Conflict with Burkina Faso lasted five days; mediated by International Court of Justice.

1991 Demonstrations against one-party rule. Moussa Traoré ousted in a coup led by Lt-Col Amadou Toumani Toure. New multiparty constitution agreed, subject to referendum.

1992 Referendum endorsed new democratic constitution. Alliance for Democracy in Mali (ADEMA) won multiparty elections; Alpha Oumar Konare elected president.

Malta
Republic of
(*Repubblika
Ta'Malta*)

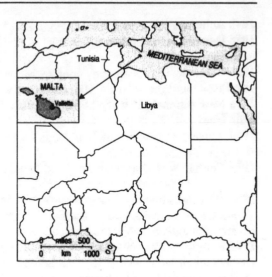

area 320 sq km/124 sq mi
capital Valletta
towns Rabat, Marsaxlokk
physical includes islands of Gozo 67 sq km/26 sq mi and Comino
2.5 sq km/1 sq mi
head of state Vincent Tabone from 1989
head of government Edward Fenech Adami from 1987
political system liberal democracy
exports vegetables, knitwear, handmade lace, plastics, electronic
equipment
currency Maltese lira
population (1990 est) 373,000; growth rate 0.7% p.a.
life expectancy men 72, women 77 (1987)
languages Maltese, English
religion Roman Catholic 98%
literacy 90% (1988)
GNP $1.6 bn; $4,750 per head (1988)
chronology
1814 Annexed to Britain by the Treaty of Paris.

1947 Achieved self-government.

1955 Dom Mintoff of the Malta Labour Party (MLP) became prime minister.

1956 Referendum approved MLP's proposal for integration with the UK. Proposal opposed by the Nationalist Party.

1958 MLP rejected the British integration proposal.

1962 Nationalists elected, with Borg Olivier as prime minister.

1964 Independence achieved from Britain, within the Commonwealth. Ten-year defence and economic-aid treaty with UK signed.

1971 Mintoff re-elected. 1964 treaty declared invalid and negotiations began for leasing the NATO base in Malta.

1972 Seven-year NATO agreement signed.

1974 Became a republic.

1979 British military base closed.

1984 Mintoff retired and was replaced by Mifsud Bonnici as prime minister and MLP leader.

1987 Edward Fenech Adami (Nationalist) elected prime minister.

1989 Vincent Tabone elected president. USA–USSR summit held offshore.

1990 Formal application made for EC membership.

1992 Nationalist Party returned to power in general election.

Mauritania
Islamic Republic of
(*République
Islamique de
Mauritanie*)

area 1,030,700 sq km/397,850 sq mi
capital Nouakchott
towns Nouadhibou, Kaédi, Zouérate
physical valley of river Senegal in S; remainder arid and flat
head of state and government Maaouia Ould Sid Ahmed Taya from
1984
political system emergent democratic republic
exports iron ore, fish, gypsum
currency ouguiya
population (1990 est) 2,038,000 (Arab-Berber 30%, black African
30%, Haratine 30%—descendants of black slaves, who remained
slaves until 1980); growth rate 3% p.a.
life expectancy men 43, women 48 (1989)
languages French (official), Hasaniya Arabic, black African languages
religion Sunni Muslim 99%
literacy 17% (1987)
GNP $843 million; $480 per head (1988)
chronology
1903 Became a French protectorate.
1960 Independence achieved from France, with Moktar Ould Daddah
as president.

1975 Western Sahara ceded by Spain. Mauritania occupied the
southern area and Morocco the north. Polisario Front formed in Sahara
to resist the occupation by Mauritania and Morocco.
1978 Daddah deposed in bloodless coup; replaced by Mohamed
Khouna Ould Haidalla. Peace agreed with Polisario Front.
1981 Diplomatic relations with Morocco broken.
1984 Haidalla overthrown by Maaouia Ould Sid Ahmed Taya.
Polisario regime formally recognized.
1985 Relations with Morocco restored.
1989 Violent clashes between Mauritanians and Senegalese. Arab-
dominated government expelled thousands of Africans into N Senegal;
governments had earlier agreed to repatriate each other's citizens
(about 250,000).
1991 Amnesty for political prisoners. Multiparty elections promised.
Calls for resignation of President Taya.
1992 First multiparty elections won by ruling Democratic and Social
Republican Party (PRDS). Diplomatic relations with Senegal resumed.

Mauritius
Republic of

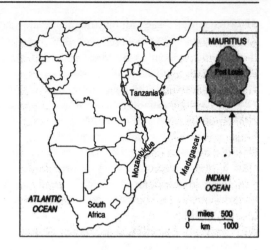

area 1,865 sq km/720 sq mi; the island of Rodrigues is part of Mauritius; there are several small island dependencies
capital Port Louis
towns Beau Bassin-Rose Hill, Curepipe, Quatre Bornes
physical mountainous, volcanic island surrounded by coral reefs
interim head of state Veerasamy Ringadoo from 1992
head of government Aneerood Jugnauth from 1982
political system liberal democratic republic
exports sugar, knitted goods, tea
currency Mauritius rupee
population (1990 est) 1,141,900, 68% of Indian origin; growth rate 1.5% p.a.
life expectancy men 64, women 71 (1989)
languages English (official), French, creole, Indian languages
religions Hindu 51%, Christian 30%, Muslim 17%
literacy 94% (1989)
GNP $1.4 bn (1987); $1,810 per head (1988)
chronology
1814 Annexed to Britain by the Treaty of Paris.
1968 Independence achieved from Britain within the Commonwealth, with Seewoosagur Ramgoolam as prime minister.

1982 Aneerood Jugnauth became prime minister.
1983 Jugnauth formed a new party, the Mauritius Socialist Movement.
Ramgoolam appointed governor general. Jugnauth formed a new
coalition government.
1985 Ramgoolam died; succeeded by Veersamy Ringadoo.
1987 Jugnauth's coalition re-elected.
1990 Attempt to create a republic failed.
1991 Jugnauth's ruling Mauritius Socialist Movement–Mauritius
Militant Movement–Rodriguais People's Organization coalition won
general election; pledge to secure republican status by 1992.
1992 Mauritius became a republic while remaining a member of the
Commonwealth. Ringadoo became interim president.

Mexico
United States of
(*Estados Unidos
Mexicanos*)

area 1,958,201 sq km/756,198 sq mi
capital Mexico City
towns Guadalajara, Monterrey, Veracruz
physical partly arid central highlands; Sierra Madre mountain ranges
E and W; tropical coastal plains
environment during the 1980s, smog levels in Mexico City exceeded
World Health Organization standards on more than 300 days of the
year. Air is polluted by 130,000 factories and 2.5 million vehicles
head of state and government Carlos Salinas de Gortari from 1988
political system federal democratic republic
exports silver, gold, lead, uranium, oil, natural gas, handicrafts, fish,
shellfish, fruits and vegetables, cotton, machinery
currency peso
population (1990 est) 88,335,000 (mixed descent 60%, Indian 30%,
Spanish descent 10%); 50% under 20 years of age; growth rate
2.6% p.a.
life expectancy men 67, women 73
languages Spanish (official) 92%, Nahuatl, Maya, Mixtec
religion Roman Catholic 97%

literacy men 92%, women 88% (1989)

GNP $126 bn (1987); $2,082 per head

chronology

1821 Independence achieved from Spain.

1846–48 Mexico at war with USA; loss of territory.

1848 Maya Indian revolt suppressed.

1864–67 Maximilian of Austria was emperor of Mexico.

1917 New constitution introduced, designed to establish permanent democracy.

1983–84 Financial crisis.

1985 Institutional Revolutionary Party (PRI) returned to power. Earthquake in Mexico City.

1986 International Monetary Fund loan agreement signed to keep the country solvent until at least 1988.

1988 PRI candidate Carlos Salinas de Gortari elected president. Debt reduction accords negotiated with USA.

1991 PRI won general election. President Salinas promised constitutional reforms.

1992 Public outrage following Guadalajara gas-explosion disaster in which 194 people died and 1,400 were injured.

Moldova
Republic of

area 33,700 sq km/13,012 sq mi
capital Chişinău (Kishinev)
towns Tiraspol, Beltsy, Bendery
physical hilly land lying largely between the rivers Prut and Dniester;
northern Moldova comprises the level plain of the Beltsy Steppe and
uplands; the climate is warm and moderately continental
head of state Mircea Snegur from 1989
head of government Valerin Murovsky from 1992
political system emergent democracy
products wine, tobacco, canned goods
population (1990) 4,400,000 (Moldavian 64%, Ukrainian 14%,
Russian 13%, Gagauzi 4%, Bulgarian 2%)
language Moldavian, allied to Romanian
religion Russian Orthodox
chronology
1940 Bessarabia in the E became part of the Soviet Union whereas the
W part remained in Romania.
1941 Bessarabia taken over by Romania–Germany.
1944 Red army reconquered Bessarabia.
1946–47 Widespread famine.

1988 A popular front, the Democratic Movement for Perestroika, campaigned for accelerated political reform.

1989 Jan–Feb: nationalist demonstrations in Chisinau. May: Moldavian Popular Front established. July: Mircea Snegur became head of state. Aug: Moldavian language granted official status triggering clashes between ethnic Russians and Moldavians. Nov: Gagauz-Khalky People's Movement formed to campaign for Gagauz autonomy.

1990 Feb: Popular Front polled strongly in supreme soviet elections. June: economic and political sovereignty declared; renamed Republic of Moldova. Oct: Gagauzi held unauthorized elections to independent parliament; state of emergency declared after inter-ethnic clashes. Trans-Dniester region declared its sovereignty. Nov: state of emergency declared in Trans-Dniester region after inter-ethnic killings.

1991 March: Moldova boycotted the USSR's constitutional referendum. Aug: independence declared after abortive anti-Gorbachev coup; Communist Party outlawed. Dec: Moldova joined new Commonwealth of Independent States (CIS).

1992 Jan: admitted into the Conference on Security and Cooperation in Europe (CSCE). Possible union with Romania discussed. March: state of emergency imposed; admitted into United Nations; diplomatic recognition granted by USA. May: further meeting on unification with Romania; Trans-Dniester region fighting intensified. July: Andrei Sangheli became premier; Moldova agreed to outside peacekeeping force. Aug: talks between Moldova and Russia began; Russian peacekeeping force reportedly deployed in Trans-Dniester region.

Monaco
Principality of

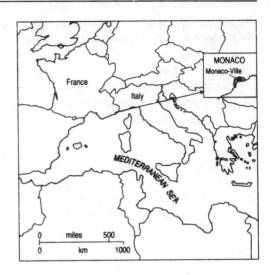

area 1.95 sq km/0.75 sq mi
capital Monaco-Ville
towns Monte Carlo, La Condamine, Fontvieille
physical steep and rugged; surrounded landwards by French territory; being expanded by filling in the sea
features aquarium and oceanographic centre; Monte Carlo film festival, motor races and casinos; world's second smallest state
head of state Prince Rainier III from 1949
head of government Jean Ausseil from 1986
political system constitutional monarchy under French protectorate
exports some light industry; economy dependent on tourism and gambling
currency French franc
population (1989) 29,000; growth rate –0.5% p.a.
languages French (official), English, Italian
religion Roman Catholic 95%
literacy 99% (1985)
chronology
1861 Became an independent state under French protection.

1918 France given a veto over succession to the throne.
1949 Prince Rainier III ascended the throne.
1956 Prince Rainier married US actress Grace Kelly.
1958 Birth of male heir, Prince Albert.
1959 Constitution of 1911 suspended.
1962 New constitution adopted.

Mongolia

State of
(*Outer Mongolia*
until 1924;
*People's Republic of
Mongolia* until
1991)

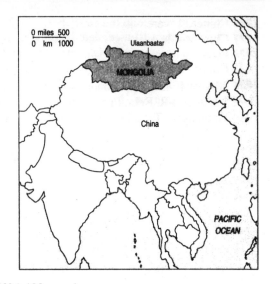

area 1,565,000 sq km/604,480 sq mi
capital Ulaanbaatar
towns Darhan, Choybalsan
physical high plateau with desert and steppe (grasslands)
head of state Punsalmaagiyn Ochirbat from 1990
head of government Puntsagiyn Jasray from 1992
political system emergent democracy
exports meat and hides, minerals, wool, livestock, grain, cement,
timber
currency tugrik
population (1990 est) 2,185,000; growth rate 2.8% p.a.
life expectancy men 63, women 67 (1989)
languages Khalkha Mongolian (official), Chinese, Russian, and
Turkic languages
religion officially none (Tibetan Buddhist Lamaism suppressed 1930s)
literacy 89% (1985)
GNP $3.6 bn; $1,820 per head (1986)
chronology
1911 Outer Mongolia gained autonomy from China.

1915 Chinese sovereignty reasserted.

1921 Chinese rule overthrown with Soviet help.

1924 People's Republic proclaimed.

1946 China recognized Mongolia's independence.

1966 20-year friendship, cooperation, and mutual-assistance pact signed with USSR. Relations with China deteriorated.

1984 Yumjaagiyn Tsedenbal, effective leader, deposed and replaced by Jambyn Batmonh.

1987 Soviet troops reduced; Mongolia's external contacts broadened.

1989 Further Soviet troop reductions.

1990 Democratization campaign launched by Mongolian Democratic Union. Punsalmaagiyn Ochirbat's Mongolian People's Revolutionary Party (MPRP) elected in free multiparty elections. Mongolian script readopted.

1991 Massive privatization programme launched as part of move towards a market economy. The word 'Republic' dropped from country's name. GDP declined by 10%.

1992 Jan: New constitution introduced. Economic situation worsened; GDP again declined by 10%. Prime minister Dashiyn Byambasuren's resignation refused. July: Puntsagiyn Jasray appointed new prime minister.

Morocco
Kingdom of
(*al-Mamlaka
al-Maghrebia*)

area 458,730 sq km/177,070 sq mi (excluding Western Sahara)
capital Rabat
towns Marrakesh, Fez, Meknès, Casablanca, Tangier, Agadir
physical mountain ranges NE–SW; fertile coastal plains in W
head of state Hassan II from 1961
head of government Mohamed Lamrani from 1992
political system constitutional monarchy
exports dates, figs, cork, wood pulp, canned fish, phosphates
currency dirham (DH)
population (1990 est) 26,249,000; growth rate 2.5% p.a.
life expectancy men 62, women 65 (1989)
languages Arabic (official) 75%, Berber 25%, French, Spanish
religion Sunni Muslim 99%
literacy men 45%, women 22% (1985 est)
GNP $18.7 bn; $750 per head (1988)
chronology
1912 Morocco divided into French and Spanish protectorates.
1956 Independence achieved as the Sultanate of Morocco.
1957 Sultan restyled king of Morocco.
1961 Hassan II came to the throne.

1969 Former Spanish province of Ifni returned to Morocco.
1972 Major revision of the constitution.
1975 Western Sahara ceded by Spain to Morocco and Mauritania.
1976 Guerrilla war in Western Sahara with the Polisario Front.
Sahrawi Arab Democratic Republic (SADR) established in Algiers.
Diplomatic relations between Morocco and Algeria broken.
1979 Mauritania signed a peace treaty with Polisario.
1983 Peace formula for Western Sahara proposed by the Organization of African Unity (OAU); Morocco agreed but refused to deal directly with Polisario.
1984 Hassan signed an agreement for cooperation and mutual defence with Libya.
1987 Cease-fire agreed with Polisario, but fighting continued.
1988 Diplomatic relations with Algeria restored.
1989 Diplomatic relations with Syria restored.
1992 Mohamed Lamrani appointed prime minister; new constitution approved in national referendum.

Mozambique
People's Republic of
(*República Popular
de Moçambique*)

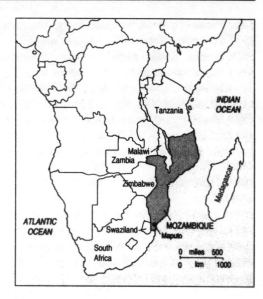

area 799,380 sq km/308,561 sq mi
capital (and chief port) Maputo
towns Beira, Nampula
physical mostly flat tropical lowland; mountains in W
head of state and government Joaquim Alberto Chissano from 1986
political system emergent democratic republic
exports prawns, cashews, sugar, cotton, tea, petroleum products, copra
currency metical (replaced escudo 1980)
population (1990 est) 14,718,000 (mainly indigenous Bantu peoples;
Portuguese 50,000); growth rate 2.8% p.a.; nearly 1 million refugees
in Malawi
life expectancy men 45, women 48 (1989)
languages Portuguese (official), 16 African languages
religion animist 60%, Roman Catholic 18%, Muslim 16%
literacy men 55%, women 22% (1985 est)
GDP $4.7 bn; $319 per head (1987)
chronology
1505 Mozambique became a Portuguese colony.

1962 Frelimo (liberation front) established.

1975 Independence achieved from Portugal as a socialist republic, with Samora Machel as president and Frelimo as the sole legal party.

1977 Renamo resistance group formed.

1983 Re-establishment of good relations with Western powers.

1984 Nkomati accord of nonaggression signed with South Africa.

1986 Machel killed in air crash; succeeded by Joaquim Chissano.

1988 Tanzania announced withdrawal of its troops. South Africa provided training for Mozambican forces.

1989 Frelimo offered to abandon Marxist-Leninism; Chissano re-elected. Renamo continued attacks on government facilities and civilians.

1990 One-party rule officially ended. Partial cease-fire agreed.

1991 Peace talks resumed in Rome, delaying democratic process. Attempted antigovernment coup thwarted.

1992 Aug: peace accord agreed upon, but fighting continued. Oct: peace accord signed, but awaited ratification by government.

Myanmar
Union of
(*Thammada
Myanmar
Naingngandaw*)
(formerly *Burma*,
until 1989)

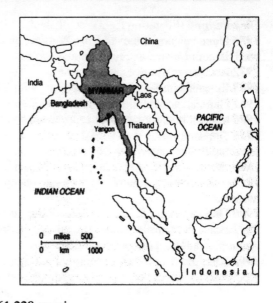

area 676,577 sq km/261,228 sq mi
capital (and chief port) Yangon (formerly Rangoon)
towns Mandalay, Moulmein, Pegu
physical over half is rainforest; rivers Irrawaddy and Chindwin in
central lowlands ringed by mountains in N, W, and E
environment landslides and flooding during the rainy season (June–
Sept) are becoming more frequent as a result of deforestation
head of state and government Than Shwe from 1992
political system military republic
exports rice, rubber, jute, teak, jade, rubies, sapphires
currency kyat
population (1990 est) 41,279,000; growth rate 1.9% p.a. (includes
Shan, Karen, Raljome, Chinese, and Indian minorities)
life expectancy men 53, women 56 (1989)
language Burmese
religions Hinayana Buddhist 85%, animist, Christian
literacy 66% (1989)
GNP $9.3 bn (1988); $210 per head (1989)

chronology
1886 United as province of British India.
1937 Became crown colony in the British Commonwealth.
1942–45 Occupied by Japan.
1948 Independence achieved from Britain. Left the Commonwealth.
1962 General Ne Win assumed power in army coup.
1973–74 Adopted presidential-style 'civilian' constitution.
1975 Opposition National Democratic Front formed.
1986 Several thousand supporters of opposition leader Suu Kyi arrested.
1988 Government resigned after violent demonstrations. General Saw Maung seized power in military coup Sept; over 1,000 killed.
1989 Martial law declared; thousands arrested including advocates of democracy and human rights. Country renamed Myanmar and capital Yangon.
1990 Breakaway opposition group formed 'parallel government' on rebel-held territory.
1991 Martial law and human-rights abuses continued. Military offensives continued. Opposition leader, Aung San Suu Kyi, received Nobel Prize for Peace.
1992 Jan–April: Pogrom against Muslim community in Arakan province, W Myanmar, carried out with army backing. April: Saw Maung replaced by Than Shwe. Several political prisoners liberated. Sept: martial law lifted, but restrictions on political freedom remained.
1993 Constitutional convention held in Yangon to discuss adoption of proposed new constitution.

Namibia
Republic of
(formerly *South
West Africa*)

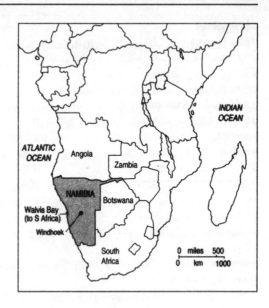

area 824,300 sq km/318,262 sq mi
capital Windhoek
towns Swakopmund, Rehoboth, Rundu
physical mainly desert
head of state Sam Nujoma from 1990
head of government Hage Geingob from 1990
political system democratic republic
exports diamonds, uranium, copper, lead, zinc
currency South African rand
population (1990 est) 1,372,000 (black African 85%, European 6%)
life expectancy blacks 40, whites 69
languages Afrikaans (spoken by 60% of white population), German,
English (all official), several indigenous languages
religion 51% Lutheran, 19% Roman Catholic, 6% Dutch Reformed
Church, 6% Anglican
literacy whites 100%, nonwhites 16%
GNP $1.6 bn; $1,300 per head (1988)

chronology

1884 German and British colonies established.

1915 German colony seized by South Africa.

1920 Administered by South Africa, under League of Nations mandate, as British South Africa.

1946 Full incorporation in South Africa refused by United Nations.

1958 South-West Africa People's Organization (SWAPO) set up to seek racial equality and full independence.

1966 South Africa's apartheid laws extended to the country.

1968 Redesignated Namibia by UN.

1978 UN Security Council Resolution 435 for the granting of full sovereignty accepted by South Africa and then rescinded.

1988 Peace talks between South Africa, Angola, and Cuba led to agreement on full independence for Namibia.

1989 Unexpected incursion by SWAPO guerrillas from Angola into Namibia threatened agreed independence. Transitional constitution created by elected representatives; SWAPO dominant party.

1990 Liberal multiparty 'independence' constitution adopted; independence achieved. Sam Nujoma elected president.

1991 Agreement on joint administration of disputed port of Walvis Bay reached with South Africa, pending final settlement of dispute.

1992 Agreement on establishment of Walvis Bay Joint Administrative Body.

Nauru
Republic of
(*Naoero*)

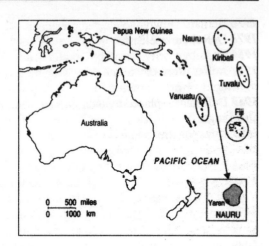

area 21 sq km/8 sq mi
capital Yaren District
physical tropical island country in SW Pacific; plateau encircled by coral cliffs and sandy beaches
head of state and government Bernard Dowiyogo from 1989
political system liberal democracy
exports phosphates
currency Australian dollar
population (1990 est) 8,100 (mainly Polynesian; Chinese 8%, European 8%); growth rate 1.7% p.a.
languages Nauruan (official), English
religion Protestant 66%, Roman Catholic 33%
literacy 99% (1988)
GNP $160 million (1986); $9,091 per head (1985)
chronology
1888 Annexed by Germany.
1920 Administered by Australia, New Zealand, and UK until independence, except 1942–45, when it was occupied by Japan.
1968 Independence achieved from Australia, New Zealand, and UK with 'special member' Commonwealth status. Hammer DeRoburt elected president.

1976 Bernard Dowiyogo elected president.
1978 DeRoburt re-elected.
1986 DeRoburt briefly replaced as president by Kennan Adeang.
1987 DeRoburt re-elected; Adeang established the Democratic Party of Nauru.
1989 DeRoburt replaced by Kensas Aroi, who was later succeeded by Dowiyogo.
1992 Dowiyogo re-elected.

Nepal
Kingdom of
(*Nepal Adhirajya*)

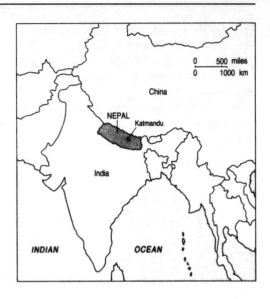

area 147,181 sq km/56,850 sq mi
capital Katmandu
towns Pátan, Moráng, Bhádgáon
physical descends from the Himalayan mountain range in N through foothills to the river Ganges plain in S
environment Nepal attracts 270,000 tourists, trekkers, and mountaineers each year. An estimated 500 kg/1,100 lb of rubbish is left by each expedition trekking or climbing in the Himalayas. Since 1952 the foothills of the Himalayas have been stripped of 40% of their forest cover
head of state King Birendra Bir Bikram Shah Dev from 1972
head of government Girija Prasad Koirala from 1991
political system constitutional monarchy
exports jute, rice, timber, oilseed
currency Nepalese rupee
population (1990 est) 19,158,000 (mainly known by name of predominant clan, the Gurkhas; the Sherpas are a Buddhist minority of

NE Nepal); growth rate 2.3% p.a.
life expectancy men 50, women 49 (1989)
language Nepali (official); 20 dialects spoken
religion Hindu 90%; Buddhist, Muslim, Christian
literacy men 39%, women 12% (1985 est)
GNP $3.1 bn (1988); $160 per head (1986)
chronology
1768 Nepal emerged as unified kingdom.
1815–16 Anglo-Nepali 'Gurkha War'; Nepal became a British-dependent buffer state.
1846–1951 Ruled by the Rana family.
1923 Independence achieved from Britain.
1951 Monarchy restored.
1959 Constitution created elected legislature.
1960–61 Parliament dissolved by king; political parties banned.
1980 Constitutional referendum held following popular agitation.
1981 Direct elections held to national assembly.
1983 Overthrow of monarch-supported prime minister.
1986 New assembly elections returned a majority opposed to *panchayat* system of partyless government.
1988 Strict curbs placed on opposition activity; over 100 supporters of banned opposition party arrested; censorship imposed.
1989 Border blockade imposed by India in treaty dispute.
1990 Panchayat system collapsed after mass prodemocracy demonstrations; new constitution introduced; elections set for May 1991.
1991 Nepali Congress Party, led by Girija Prasad Koirala, won the general election.
1992 Communists led anti-government demonstrations in Katmandu and Pátan.

Netherlands
Kingdom of the
(*Koninkrijk der
Nederlanden*),
popularly referred to
as *Holland*

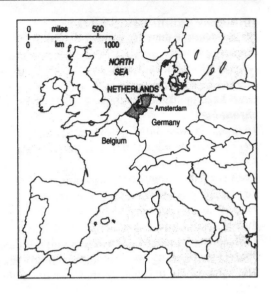

area 41,863 sq km/16,169 sq mi
capital Amsterdam
towns The Hague, Utrecht, Eindhoven, Maastricht, Rotterdam
physical flat coastal lowland; rivers Rhine, Scheldt, Maas; Frisian
Islands
territories Aruba, Netherlands Antilles (Caribbean)
environment the country lies at the mouths of three of Europe's most
polluted rivers, the Maas, Rhine, and Scheldt. Dutch farmers
contribute to this pollution by using the world's highest quantities of
nitrogen-based fertilizer per hectare/acre per year
head of state Queen Beatrix Wilhelmina Armgard from 1980
head of government Ruud Lubbers from 1989
political system constitutional monarchy
exports dairy products, flower bulbs, vegetables, petrochemicals,
electronics
currency guilder
population (1990 est) 14,864,000 (including 300,000 of Dutch-
Indonesian origin absorbed 1949–64 from former colonial

possessions); growth rate 0.4% p.a.
life expectancy men 74, women 81 (1989)
language Dutch
religions Roman Catholic 40%, Protestant 31%
literacy 99% (1989)
GNP $223 bn (1988); $13,065 per head (1987)
chronology
1940–45 Occupied by Germany during World War II.
1947 Joined Benelux customs union.
1948 Queen Juliana succeeded Queen Wilhelmina to the throne.
1949 Became a founding member of North Atlantic Treaty
Organization (NATO).
1953 Dykes breached by storm; nearly 2,000 people and tens of
thousands of cattle died in flood.
1958 Joined European Economic Community.
1980 Queen Juliana abdicated in favour of her daughter Beatrix.
1981 Opposition to cruise missiles averted their being sited on Dutch
soil.
1989 Prime Minister Ruud Lubbers resigned; new Lubbers-led
coalition elected.
1991 Treaty on political and monetary union signed by European
Community (EC) members at Maastricht.
1992 Maastricht Treaty ratified.

New Zealand
Dominion of

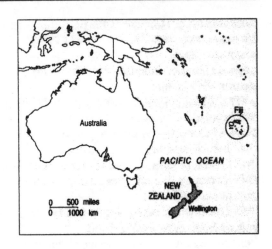

area 268,680 sq km/103,777 sq mi
capital Wellington
towns Hamilton, Palmerston North, Christchurch, Dunedin, Auckland
physical comprises North Island, South Island, Stewart Island,
Chatham Islands, and minor islands; mainly mountainous
overseas territories Tokelau (three atolls transferred 1926 from former
Gilbert and Ellice Islands colony); Niue Island (one of the Cook
Islands, separately administered from 1903: chief town Alafi); Cook
Islands are internally self-governing but share common citizenship
with New Zealand; Ross Dependency in Antarctica
head of state Elizabeth II from 1952, represented by governor general
(Catherine Tizard from 1990)
head of government Jim Bolger from 1990
political system constitutional monarchy
exports lamb, beef, wool, leather, dairy products, processed foods,
kiwi fruit, seeds and breeding stock, timber, paper, pulp, light aircraft
currency New Zealand dollar
population (1990 est) 3,397,000 (European, mostly British, 87%;
Polynesian, mostly Maori, 12%); growth rate 0.9% p.a.
life expectancy men 72, women 78 (1989)
languages English (official), Maori

religions Protestant 50%, Roman Catholic 15%
literacy 99% (1989)
GNP $37 bn; $11,040 per head (1988)
chronology
1840 New Zealand became a British colony.
1907 Created a dominion of the British Empire.
1931 Granted independence from Britain.
1947 Independence within the Commonwealth confirmed by the New Zealand parliament.
1972 National Party government replaced by Labour Party, with Norman Kirk as prime minister.
1974 Kirk died; replaced by Wallace Rowling.
1975 National Party returned, with Robert Muldoon as prime minister.
1984 Labour Party returned under David Lange.
1985 Non-nuclear military policy created disagreements with France and the USA.
1987 National Party declared support for the Labour government's non-nuclear policy. Lange re-elected. New Zealand officially classified as a 'friendly' rather than 'allied' country by the USA because of its non-nuclear military policy.
1988 Free-trade agreement with Australia signed.
1989 Lange resigned over economic differences with finance minister (he cited health reasons); replaced by Geoffrey Palmer.
1990 Palmer replaced by Mike Moore. Labour Party defeated by National Party in general election; Jim Bolger became prime minister.
1991 Formation of amalgamated Alliance Party set to challenge two-party system.
1992 Ban on visits by US warships lifted. Constitutional change agreed upon.

Nicaragua
Republic of
(*República de*
Nicaragua)

area 127,849 sq km/49,363 sq mi
capital Managua
towns León, Granada, Corinto, Puerto Cabezas, El Bluff
physical narrow Pacific coastal plain separated from broad Atlantic
coastal plain by volcanic mountains and lakes Managua and Nicaragua
head of state and government Violeta Barrios de Chamorro from 1990
political system emergent democracy
exports coffee, cotton, sugar, bananas, meat
currency cordoba
population (1990) 3,606,000 (mestizo 70%, Spanish descent 15%,
Indian or black 10%); growth rate 3.3% p.a.
life expectancy men 61, women 63 (1989)
languages Spanish (official), Indian, English
religion Roman Catholic 95%
literacy 66% (1986)
GNP $2.1 bn; $610 per head (1988)
chronology
1838 Independence achieved from Spain.
1926–1933 Occupied by US marines.
1936 General Anastasio Somoza elected president; start of

near-dictatorial rule by Somoza family.

1962 Sandinista National Liberation Front (FSLN) formed to fight Somoza regime.

1979 Somoza government ousted by FSLN.

1982 Subversive activity against the government by right-wing Contra guerrillas promoted by the USA. State of emergency declared.

1984 The USA mined Nicaraguan harbours.

1985 Denunciation of Sandinista government by US president Reagan. FSLN won assembly elections.

1987 Central American peace agreement cosigned by Nicaraguan leaders.

1988 Peace agreement failed. Nicaragua held talks with Contra rebel leaders. Hurricane left 180,000 people homeless.

1989 Demobilization of rebels and release of former Somozan supporters; cease-fire ended.

1990 FSLN defeated by UNO, a US-backed coalition; Violeta Barrios de Chamorro elected president. Antigovernment riots.

1991 First presidential state visit to USA for over fifty years.

1992 June: US aid suspended because of concern over role of Sandinista in Nicaraguan government. Sept: around 16,000 made homeless by earthquake.

Niger
Republic of
(*République du
Niger*)

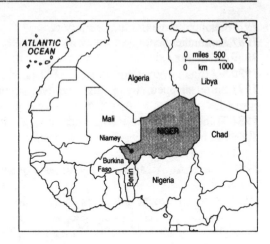

area 1,186,408 sq km/457,953 sq mi
capital Niamey
towns Zinder, Maradi, Tahoua
physical desert plains between hills in N and savanna in S; river Niger
in SW, Lake Chad in SE
head of state Ali Saibu from 1987
head of government to be elected
political system military republic
exports peanuts, livestock, gum arabic, uranium
currency franc CFA
population (1990 est) 7,691,000; growth rate 2.8% p.a.
life expectancy men 48, women 50 (1989)
languages French (official), Hausa, Djerma, and other minority
languages
religions Sunni Muslim 85%, animist 15%
literacy men 19%, women 9% (1985 est)
GNP $2.2 bn; $310 per head (1987)
chronology
1960 Achieved full independence from France; Hamani Diori elected
first president.
1974 Diori ousted in army coup led by Seyni Kountché.

1977 Cooperation agreement signed with France.
1987 Kountché died; replaced by Col Ali Saibu.
1989 Saibu elected president without opposition.
1990 Multiparty politics promised.
1991 Saibu stripped of executive powers; transitional government formed.
1992 Transitional government collapsed. Constitutional change allowing for multiparty politics approved in referendum.

Nigeria
Federal Republic of

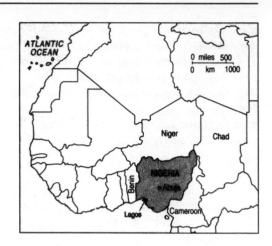

area 923,773 sq km/356,576 sq mi
capital Abuja
towns Ibadan, Ogbomosho, Kano, Lagos, Port Harcourt, Warri
physical arid savanna in N; tropical rainforest in S, with mangrove
swamps along the coast; river Niger forms wide delta; mountains
in SE
environment toxic waste from northern industrialized countries has
been dumped in Nigeria
head of state and government Ibrahim Babangida from 1985
political system military republic pending promised elections
exports petroleum (largest oil resources in Africa), cocoa, peanuts,
palm oil (Africa's largest producer), cotton, rubber, tin
currency naira
population (1991) 88,514,500 (Yoruba in W, Ibo in E, and Hausa–
Fulani in N); growth rate 3.3% p.a.
life expectancy men 47, women 49 (1989)
languages English (official), Hausa, Ibo, Yoruba
religions Sunni Muslim 50% (in N), Christian 40% (in S), local
religions 10%
literacy men 54%, women 31% (1985 est)
GNP $78 bn (1987); $790 per head (1984)

chronology

1914 N Nigeria and S Nigeria united to become Britain's largest African colony.

1954 Nigeria became a federation.

1960 Independence achieved from Britain within the Commonwealth.

1963 Became a republic, with Nnamdi Azikiwe as president.

1966 Military coup, followed by a counter-coup led by General Yakubu Gowon. Slaughter of many members of the Ibo tribe in north.

1967 Conflict over oil revenues led to declaration of an independent Ibo state of Biafra and outbreak of civil war.

1970 Surrender of Biafra and end of civil war.

1975 Gowon ousted in military coup; second coup put General Olusegun Obasanjo in power.

1979 Shehu Shagari became civilian president.

1983 Shagari's government overthrown in coup by Maj-Gen Muhammadu Buhari.

1985 Buhari replaced in a bloodless coup led by Maj-General Ibrahim Babangida.

1989 Two new political parties approved. Babangida promised a return to pluralist politics; date set for 1992.

1991 Nine new states created. Babangida confirmed his commitment to democratic rule for 1992.

1992 Multiparty elections won by Babangida's Social Democratic Party (SDP). Primary elections to be introduced; transition to civilian rule delayed.

Norway
Kingdom of
(*Kongeriket Norge*)

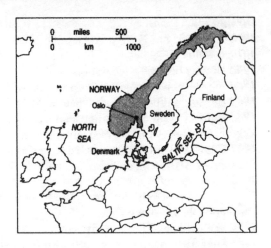

area 387,000 sq km/149,421 sq mi (includes Svalbard and Jan Mayen)
capital Oslo
towns Bergen, Trondheim, Stavanger
physical mountainous with fertile valleys and deeply indented coast;
forests cover 25%; extends N of Arctic Circle
territories dependencies in the Arctic (Svalbard and Jan Mayen) and in
Antarctica (Bouvet and Peter I Island, and Queen Maud Land)
environment an estimated 80% of the lakes and streams in the
southern half of the country have been severely acidified by acid rain
head of state Harald V from 1991
head of government Gro Harlem Brundtland from 1990
political system constitutional monarchy
exports petrochemicals from North Sea oil and gas, paper, wood pulp,
furniture, iron ore and other minerals, high-tech goods, fish
currency krone
population (1990 est) 4,214,000; growth rate 0.3% p.a.
life expectancy men 73, women 80 (1989)
languages Norwegian (official); there are Saami- (Lapp-) and Finnish-
speaking minorities
religion Evangelical Lutheran (endowed by state) 94%
literacy 100% (1989)

GNP $89 bn (1988); $13,790 per head (1984)

chronology

1814 Became independent from Denmark; ceded to Sweden.

1905 Links with Sweden ended; full independence achieved.

1940–45 Occupied by Germany.

1949 Joined North Atlantic Treaty Organization (NATO).

1952 Joined Nordic Council.

1957 King Haakon VII succeeded by his son Olaf V.

1960 Joined European Free Trade Association (EFTA).

1972 Accepted into membership of European Economic Community; application withdrawn after a referendum.

1988 Gro Harlem Brundtland awarded Third World Prize.

1989 Jan P Syse became prime minister.

1990 Brundtland returned to power.

1991 King Olaf V died; succeeded by his son Harald V.

1992 Defied whaling ban to resume whaling industry. Brundtland relinquished leadership of the Labour Party. Formal application made for EC membership.

Oman
Sultanate of
(*Saltanat 'Uman*)

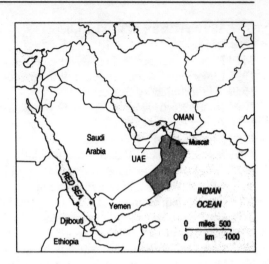

area 272,000 sq km/105,000 sq mi
capital Muscat
towns Salalah, Nizwa
physical mountains to N and S of a high arid plateau; fertile coastal
strips NW and S
head of state and government Qaboos bin Said from 1970
political system absolute monarchy
exports oil, dates, silverware, copper
currency rial Omani
population (1990 est) 1,305,000; growth rate 3.0% p.a.
life expectancy men 55, women 58 (1989)
languages Arabic (official), English, Urdu, other Indian languages
religions Ibadhi Muslim 75%, Sunni Muslim, Shi'ite Muslim, Hindu
literacy 20% (1989)
GNP $7.5 bn (1987); $5,070 per head (1988)
chronology
1951 The Sultanate of Muscat and Oman achieved full independence
from Britain. Treaty of Friendship with Britain signed.
1970 After 38 years' rule, Sultan Said bin Taimur replaced in coup by
his son Qaboos bin Said. Name changed to Sultanate of Oman.

1975 Left-wing rebels in south defeated.

1982 Memorandum of Understanding with UK signed, providing for regular consultation on international issues.

1985 Diplomatic ties established with USSR.

1991 Sent troops to Operation Desert Storm, as part of coalition opposing Iraq's occupation of Kuwait.

Pakistan
Islamic Republic of

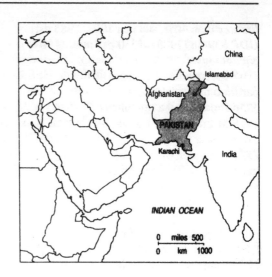

area 796,100 sq km/307,295 sq mi; one-third of Kashmir under
Pakistani control
capital Islamabad
towns Karachi, Lahore, Rawalpindi, Peshawar
physical fertile Indus plain in E, Baluchistan plateau in W, mountains
in N and NW
environment about 68% of irrigated land is waterlogged or suffering
from salinization
head of state Ghulam Ishaq Khan from 1988
head of government Nawaz Sharif from 1990
political system emergent democracy
exports cotton textiles, rice, leather, carpets
currency Pakistan rupee
population (1990 est) 113,163,000 (Punjabi 66%, Sindhi 13%);
growth rate 3.1% p.a.
life expectancy men 54, women 55 (1989)
languages Urdu and English (official); Punjabi, Sindhi, Pashto,
Baluchi, other local dialects
religion Sunni Muslim 75%, Shi'ite Muslim 20%, Hindu 4%

literacy men 40%, women 19% (1985 est)
GDP $39 bn (1988); $360 per head (1984)
chronology
1947 Independence achieved from Britain; Pakistan formed following partition of British India.
1956 Proclaimed a republic.
1958 Military rule imposed by General Ayub Khan.
1969 Power transferred to General Yahya Khan.
1971 Secession of East Pakistan (Bangladesh). After civil war, power transferred to Zulfiqar Ali Bhutto.
1977 Bhutto overthrown in military coup by General Zia ul-Haq; martial law imposed.
1979 Bhutto executed.
1981 Opposition Movement for the Restoration of Democracy formed. Islamization process pushed forward.
1985 Nonparty elections held; amended constitution adopted; martial law and ban on political parties lifted.
1986 Agitation for free elections launched by Benazir Bhutto.
1988 Zia introduced Islamic legal code, the Shari'a. He was killed in a military plane crash Aug. Benazir Bhutto elected prime minister Nov.
1989 Pakistan rejoined the Commonwealth.
1990 Army mobilized in support of Muslim separatists in Indian Kashmir. Bhutto dismissed on charges of incompetence and corruption. Islamic Democratic Alliance (IDA), led by Nawaz Sharif, won Oct general election.
1991 Shari'a bill enacted; privatization and economic deregulation programme launched.
1992 Sept: Floods devastated north of country. Oct: Pakistan elected to UN Security Council 1993–95.

Panama
Republic of
(*República de
Panamá*)

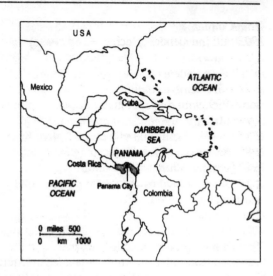

area 77,100 sq km/29,768 sq mi
capital Panamá (Panama City)
towns Cristóbal, Balboa, Colón, David
physical coastal plains and mountainous interior; tropical rainforest in
E and NW; Pearl Islands in Gulf of Panama
head of state and government Guillermo Endara from 1989
political system emergent democratic republic
exports bananas, petroleum products, copper, shrimps, sugar
currency balboa
population (1990 est) 2,423,000 (mestizo, or mixed race, 70%, West
Indian 14%, European descent 10%, Indian (Cuna, Choco, Guayami)
6%); growth rate 2.2% p.a.
life expectancy men 71, women 75 (1989)
languages Spanish (official), English
religions Roman Catholic 93%, Protestant 6%
literacy 87% (1989)
GNP $4.2 bn (1988); $1,970 per head (1984)
chronology
1821 Achieved independence from Spain; joined the confederacy of

Gran Colombia.

1903 Full independence achieved on separation from Colombia.

1974 Agreement to negotiate full transfer of the Panama Canal from the USA to Panama.

1977 USA–Panama treaties transferred the canal to Panama, effective from 1999, with the USA guaranteeing its protection and an annual payment.

1984 Nicolás Ardito Barletta elected president.

1985 Barletta resigned; replaced by Eric Arturo del Valle.

1987 General Noriega (head of the National Guard and effective ruler of Panama) resisted calls for his removal, despite suspension of US military and economic aid.

1988 Del Valle replaced by Manuel Solis Palma. Noriega, charged with drug smuggling by the USA, declared a state of emergency.

1989 Opposition won election; Noriega declared results invalid; Francisco Rodríguez sworn in as president. Coup attempt against Noriega failed; Noriega declared head of government by assembly. 'State of war' with the USA announced. US invasion deposed Noriega; Guillermo Endara installed as president; Noriega sought asylum in Vatican embassy; later surrendered and taken to US for trial.

1991 Attempted antigovernment coup foiled. Army abolished.

1992 Noriega found guilty of drug offences. Referendum voted down overwhelmingly government's constitutional changes, including abolition of a standing army.

Papua New Guinea

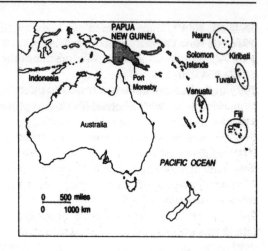

area 462,840 sq km/178,656 sq mi
capital Port Moresby (on E New Guinea)
towns Lae, Rabaul, Madang
physical mountainous; includes tropical islands of New Ireland, New Britain, and Bougainville; Admiralty Islands, D'Entrecasteaux Islands, and Louisiade Archipelago
head of state Elizabeth II, represented by governor general
head of government Paias Wingti from 1992
political system liberal democracy
exports copra, coconut oil, palm oil, tea, copper, gold, coffee
currency kina
population (1989 est) 3,613,000 (Papuans, Melanesians, Negritos, various minorities); growth rate 2.6% p.a.
life expectancy men 53, women 54 (1987)
languages English (official); pidgin English, 715 local languages
religions Protestant 63%, Roman Catholic 31%, local faiths
literacy men 55%, women 36% (1985 est)
GNP $2.5 bn; $730 per head (1987)
chronology
1883 Annexed by Queensland; became the Australian Territory of Papua.

1884 NE New Guinea annexed by Germany; SE claimed by Britain.
1914 NE New Guinea occupied by Australia.
1921–42 Held as a League of Nations mandate.
1942–45 Occupied by Japan.
1975 Independence achieved from Australia, within the
Commonwealth, with Michael Somare as prime minister.
1980 Julius Chan became prime minister.
1982 Somare returned to power.
1985 Somare challenged by deputy prime minister, Paias Wingti, who
later formed a five-party coalition government.
1988 Wingti defeated on no-confidence vote and replaced by Rabbie
Namaliu, who established a six-party coalition government.
1989 State of emergency imposed on Bougainville in response to
separatist violence.
1991 Peace accord signed with Bougainville secessionists. Economic
boom as gold production doubled. Wiwa Korowi elected as new
governor general. Deputy prime minister, Ted Diro, resigned, having
been found guilty of corruption.
1992 April: killings by outlawed Bougainville secessionists reported.
July: Wingti elected premier.

Paraguay
Republic of
(*República del
Paraguay*)

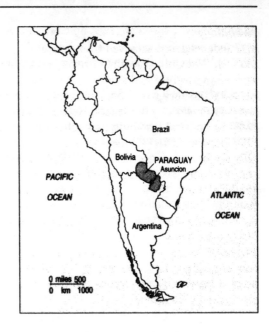

area 406,752 sq km/157,006 sq mi
capital Asunción
towns Puerto Presidente Stroessner, Pedro Juan Caballero, Concepción
physical low marshy plain and marshlands; divided by Paraguay
River; Paraná River forms SE boundary
head of state and government General Andrés Rodríguez from 1989
political system emergent democratic republic
exports cotton, soya beans, timber, vegetable oil, maté
currency guaraní
population (1990 est) 4,660,000 (95% mixed Guarani Indian–Spanish
descent); growth rate 3.0% p.a.
life expectancy men 67, women 72 (1989)
languages Spanish 6% (official), Guarani 90%
religion Roman Catholic 97%
literacy men 91%, women 85% (1985 est)
GNP $7.4 bn; $1,000 per head (1987)

chronology

1811 Independence achieved from Spain.

1865–70 War with Argentina, Brazil, and Uruguay; less than half the population survived and much territory was lost.

1932–35 Territory won from Bolivia during the Chaco War.

1940–48 Presidency of General Higinio Morínigo.

1948–54 Political instability; six different presidents.

1954 General Alfredo Stroessner seized power.

1989 Stroessner ousted in coup led by General Andrés Rodríguez. Rodríguez elected president; Colorado Party won the congressional elections.

1991 Colorado Party successful in assembly elections.

Peru
Republic of
(*República del Perú*)

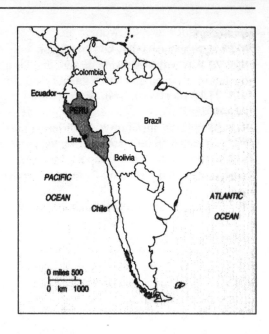

area 1,285,200 sq km/496,216 sq mi
capital Lima, including port of Callao
towns Arequipa, Iquitos, Chiclayo, Trujillo
physical Andes mountains NW–SE cover 27% of Peru, separating
Amazon rainforest in NE from coastal plain in W; desert along coast
environment an estimated 38% of the 8,000 sq km/3,100 sq mi of
coastal lands under irrigation is either waterlogged or suffering from
saline water. Only half the population has access to clean drinking
water
head of state and government Alberto Fujimori from 1990
political system democratic republic
exports coca, coffee, alpaca, llama and vicuña wool, fish meal, lead
(largest producer in South America), copper, iron, oil
currency new sol
population (1990 est) 21,904,000 (Indian, 46%; mixed Spanish–
Indian descent 43%); growth rate 2.6% p.a.

life expectancy men 61, women 66
languages Spanish 68%, Quechua 27% (both official), Aymara 3%
religion Roman Catholic 90%
literacy men 91%, women 78% (1985 est)
GNP $19.6 bn (1988); $940 per head (1984)
chronology
1824 Independence achieved from Spain.
1849–74 Some 80,000–100,000 Chinese labourers arrived in Peru to
fill menial jobs such as collecting guano.
1902 Boundary dispute with Bolivia settled.
1927 Boundary dispute with Colombia settled.
1942 Boundary dispute with Ecuador settled.
1948 Army coup, led by General Manuel Odría, installed a military
government.
1963 Return to civilian rule, with Fernando Belaúnde Terry as
president.
1968 Return of military government in a bloodless coup by General
Juan Velasco Alvarado.
1975 Velasco replaced, in a bloodless coup, by General Morales
Bermúdez.
1980 Return to civilian rule, with Fernando Belaúnde as president.
1981 Boundary dispute with Ecuador renewed.
1985 Belaúnde succeeded by Social Democrat Alan García Pérez.
1988 García pressured to seek help from the International Monetary
Fund. (IMF)
1989 Mario Vargas Llosa entered presidential race; his Democratic
Front won municipal elections Nov.
1990 Alberto Fujimori defeated Vargas Llosa in presidential elections.
Assassination attempt on president failed.
1992 April: Fujimori sided with army to avert coup, announcing
crackdown on rebels and drug traffickers. USA suspended
humanitarian aid. Sendero Luminoso ('Shining Path') terrorists
continued campaign of violence. May: Fujimori promised return to
democracy. Oct: Sendero Luminoso leader, Abimael Guzman
Reynoso, arrested received life sentence. Nov: anti-government coup
foiled; single-chamber legislature replaced two-chamber system.

Philippines
Republic of the
(*Republika ng
Pilipinas*)

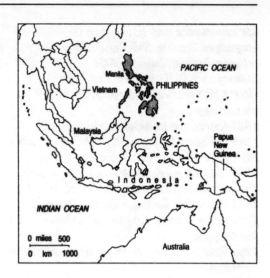

area 300,000 sq km/115,800 sq mi
capital Manila (on Luzon)
towns Quezon City (Luzon), Zamboanga (Mindanao), Cebu, Davao
(on Mindanao), Iloilo
physical comprises over 7,000 islands; volcanic mountain ranges
traverse main chain N–S; 50% still forested. The largest islands are
Luzon 108,172 sq km/41,754 sq mi and Mindanao 94,227 sq km/
36,372 sq mi; others include Samar, Negros, Palawan, Panay,
Mindoro, Leyte, Cebu, and the Sulu group
environment cleared for timber, tannin, and the creation of fish ponds,
the mangrove forest was reduced from 5,000 sq km/1,930 sq mi to
380 sq km/146 sq mi between 1920 and 1988
head of state and government Fidel Ramos from 1992
political system emergent democracy
exports sugar, copra (world's largest producer) and coconut oil, timber,
copper concentrates, electronics, clothing
currency peso
population (1990 est) 66,647,000 (93% Malaysian); growth rate
2.4% p.a.

life expectancy men 63, women 69 (1989)
languages Tagalog (Filipino, official); English and Spanish
religions Roman Catholic 84%, Protestant 9%, Muslim 5%
literacy 88% (1989)
GNP $38.2 bn; $667 per head (1988)
chronology
1542 Named the Philippines (Filipinas) by Spanish explorers.
1565 Conquered by Spain.
1898 Ceded to the USA after Spanish–American War.
1935 Granted internal self-government.
1942–45 Occupied by Japan.
1946 Independence achieved from USA.
1965 Ferdinand Marcos elected president.
1983 Opposition leader Benigno Aquino murdered by military guard.
1986 Marcos overthrown by Corazon Aquino's People's Power movement.
1987 'Freedom constitution' adopted, giving Aquino mandate to rule until June 1992; People's Power won majority in congressional elections. Attempted right-wing coup suppressed. Communist guerrillas active. Government in rightward swing.
1988 Land Reform Act gave favourable compensation to holders of large estates.
1989 Referendum on southern autonomy failed; Marcos died in exile; Aquino refused his burial in Philippines. Sixth coup attempt suppressed with US aid; Aquino declared state of emergency.
1990 Seventh coup attempt survived by President Aquino.
1991 June: eruption of Mount Pinatubo, hundreds killed. USA agreed to give up Clark Field airbase but keep Subic Bay naval base for ten more years. Sept: Philippines Senate voted to urge withdrawal of all US forces. US renewal of Subic Bay lease rejected. Nov: Imelda Marcos returned.
1992 Fidel Ramos elected to replace Aquino.

Poland
Republic of
(*Polska
Rzeczpospolita*)

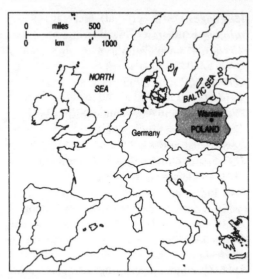

area 127,886 sq km/49,325 sq mi
capital Warsaw
towns Lódź, Kraków, Wroclaw, Poznań, Katowice, Gdańsk
physical part of the great plain of Europe; Vistula, Oder, and Neisse
rivers; Sudeten, Tatra, and Carpathian mountains on S frontier
environment atmospheric pollution derived from coal (producing 90%
of the country's electricity), toxic waste from industry, and lack of
sewage treatment have resulted in the designation of 27 ecologically
endangered areas. Half the country's lakes have been seriously
contaminated and three-quarters of its drinking water does not meet
official health standards
head of state Lech Walesa from 1990
head of government Hanna Suchocka from 1992
political system emergent democratic republic
exports coal, softwood timber, chemicals, machinery, ships, vehicles,
meat, copper (Europe's largest producer)
currency zloty
population (1990 est) 38,363,000; growth rate 0.6% p.a.
life expectancy men 66, women 74 (1989)

languages Polish (official), German
religion Roman Catholic 95%
literacy 98% (1989)
GNP $276 billion (1988); $2,000 per head (1986)
chronology
1918 Poland revived as independent republic.
1939 German invasion and occupation.
1944 Germans driven out by Soviet forces.
1945 Polish boundaries redrawn at Potsdam Conference.
1947 Communist people's republic proclaimed.
1956 Poznań riots. Wladyslav Gomulka installed as Polish United Workers' Party (PUWP) leader.
1970 Gomulka replaced by Edward Gierek after Gdańsk riots.
1980 Solidarity emerged as a free trade union following Gdańsk disturbances.
1981–83 Martial law imposed by General Wojciech Jaruzelski.
1985 Zbigniew Messner became prime minister.
1988 Solidarity-led strikes and demonstrations called off after pay increases. Messner replaced by reformist Mieczyslaw Rakowski.
1989 Solidarity relegalized. April: new 'socialist pluralist' constitution formed. June: widespread success for Solidarity in assembly elections, the first open elections in 40 years. July: Jaruzelski elected president. Sept: 'Grand coalition', first non-Communist government since World War II formed; economic restructuring undertaken on free-market lines; W Europe and US create $1 billion aid package.
1990 Jan: PUWP dissolved; replaced by Social Democratic Party and breakaway Union of Social Democrats. Lech Walesa elected president; Dec: prime minister Mazowiecki resigned.
1991 Oct: Multiparty general election produced inconclusive result. Five-party centre-right coalition formed under Jan Olszewski. Treaty signed agreeing to complete withdrawal of Soviet troops.
1992 June: Olszewski ousted on vote of no confidence; succeeded by Waldemar Pawlak. July: Hanna Suchocka replaced Pawlak as Poland's first woman prime minister.
1993 March: 14% of workforce (2.6 million) out of work.

Portugal
Republic of
(*República
Portuguesa*)

area 92,000 sq km/35,521 sq mi (including the Azores and Madeira)
capital Lisbon
towns Coimbra, Pôrto, Setúbal
physical mountainous in N, plains in S
head of state Mario Alberto Nobre Lopes Soares from 1986
head of government Aníbal Cavaco Silva from 1985
political system democratic republic
exports wine, olive oil, resin, cork, sardines, textiles, clothing, pottery, pulpwood
currency escudo
population (1990 est) 10,528,000; growth rate 0.5% p.a.
life expectancy men 71, women 78 (1989)
language Portuguese
religion Roman Catholic 97%
literacy men 89%, women 80% (1985)
GNP $33.5 bn (1987); $2,970 per head (1986)
chronology
1928–68 Military dictatorship under António de Oliveira Salazar.
1968 Salazar succeeded by Marcello Caetano.

1974 Caetano removed in military coup led by General Antonio Ribeiro de Spínola. Spínola replaced by General Francisco da Costa Gomes.

1975 African colonies became independent.

1976 New constitution, providing for return to civilian rule, adopted. Minority government appointed, led by Socialist Party leader Mario Soares.

1978 Soares resigned.

1980 Francisco Balsemão formed centre-party coalition after two years of political instability.

1982 Draft of new constitution approved, reducing powers of presidency.

1983 Centre-left coalition government formed.

1985 Aníbal Cavaco Silva became prime minister.

1986 Mario Soares elected first civilian president in 60 years. Portugal joined European Community.

1988 Portugal joined Western European Union.

1989 Constitution amended to allow major state enterprises to be denationalized.

1991 Mario Soares re-elected president; Social Democrat (PSD) majority slightly reduced in assembly elections.

Qatar
State of
(*Dawlat Qatar*)

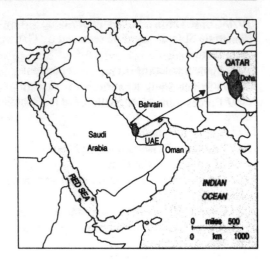

area 11,400 sq km/4,402 sq mi
capital Doha
town Dukhan
physical mostly flat desert with salt flats in S
head of state and government Sheik Khalifa bin Hamad al-Thani
from 1972
political system absolute monarchy
exports oil, natural gas, petrochemicals, fertilizers, iron, steel
currency riyal
population (1990 est) 498,000 (half in Doha; Arab 40%, Indian 18%,
Pakistani 18%); growth rate 3.7% p.a.
life expectancy men 68, women 72 (1989)
languages Arabic (official), English
religion Sunni Muslim 95%
literacy 60% (1987)
GNP $5.9 bn (1983); $35,000 per head
chronology
1916 Qatar became a British protectorate.
1939 Oil was discovered in W Qatar.
1949 Qatar began exporting oil.

1970 Constitution adopted, confirming the emirate as an absolute
monarchy
1971 Independence achieved from Britain
1972 Emir Sheik Ahmad replaced in bloodless coup by his cousin,
Crown Prince Sheik Khalifa
1991 Forces joined UN coalition in Gulf War against Iraq.

Romania

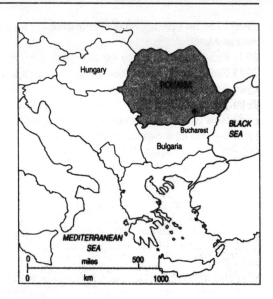

area 237,500 sq km/91,699 sq mi
capital Bucharest
towns Braşov, Timişoara, Cluj–Napoca, Iaşi, Galaţi, Constanta, Braila
physical mountains surrounding a plateau, with river plains S and E
environment although sulphur-dioxide levels are low, only 20% of the country's rivers can provide drinkable water
head of state Ion Iliescu from 1989
head of government Theodor Stolojan from 1991
political system emergent democratic republic
exports petroleum products and oilfield equipment, electrical goods, cars, cereals
currency leu
population (1990 est) 23,269,000 (Romanians 89%, Hungarians 7.9%, Germans 1.6%); growth rate 0.5% p.a.
life expectancy men 67, women 73 (1989)
languages Romanian (official), Hungarian, German
religions Romanian Orthodox 80%, Roman Catholic 6%
literacy 98% (1988)

GNP $151 bn (1988); $6,400 per head
recent chronology
1944 Pro-Nazi Antonescu government overthrown.
1945 Communist-dominated government appointed.
1947 Boundaries redrawn. King Michael abdicated and People's Republic proclaimed.
1949 New Soviet-style constitution adopted. Joined Comecon.
1952 Second new Soviet-style constitution.
1955 Romania joined Warsaw Pact.
1958 Soviet occupation forces removed.
1965 New constitution adopted.
1974 Ceauşescu created president.
1985–86 Winters of austerity and power cuts.
1987 Workers demonstrated against austerity programme.
1988–89 Relations with Hungary deteriorated over 'systematization programme'.
1989 Announcement that all foreign debt paid off. Razing of villages and building of monuments to Ceauşescu. Communist orthodoxy reaffirmed; demonstrations violently suppressed; massacre in Timisoara. Army joined uprising; heavy fighting; bloody overthrow of Ceauşescu regime in 'Christmas Revolution'; Ceauşescu and wife tried and executed; estimated 10,000 dead in civil warfare. Power assumed by new National Salvation Front, headed by Ion Iliescu.
1990 Securitate secret police replaced by new Romanian Intelligence Service (RIS); religious practices resumed; mounting strikes and protests against effects of market economy.
1991 April: treaty on cooperation and good neighbourliness signed with USSR. Aug: privatization law passed. Sept: prime minister Petre Roman resigned following riots; succeeded by Theodor Stolojan heading a new cross-party coalition government. Dec: new constitution endorsed by referendum.
1992 Iliescu re-elected in presidential runoff.

Russian Federation
formerly (until 1991)
Russian Soviet
Federal Socialist
Republic (RSFSR)

area 17,075,500 sq km/6,591,100 sq mi
capital Moscow
towns St Petersburg (Leningrad), Nizhny-Novgorod (Gorky), Rostov-on-Don, Samara (Kuibyshev), Tver (Kalinin), Volgograd
physical fertile Black Earth district; extensive forests; the Ural Mountains with large mineral resources
head of state Boris Yeltsin from 1990/91
head of government Viktor Chernomyrdin from 1992
political system emergent democracy
products iron ore, coal, oil, gold, platinum, and other minerals, agricultural produce
currency rouble
population (1990) 148,000,000 (82% Russian, Tatar 4%, Ukrainian 3%, Chuvash 1%)
language Great Russian
religion traditionally Russian Orthodox
recent chronology
1945 Became a founding member of United Nations.
1988 Aug: Democratic Union formed in Moscow as political party

opposed to totalitarianism. Oct: Russian-language demonstrations in Leningrad; Tsarist flag raised.

1989 March: Boris Yeltsin elected to USSR Congress of People's Deputies. Sept: conservative-nationalist Russian United Workers' Front established in Sverdlovsk.

1990 May: Yeltsin narrowly elected RSFSR president by Russian parliament. June: economic and political sovereignty declared; Ivan Silaev became Russian prime minister. July: Yeltsin resigned his party membership. Dec: rationing introduced in some cities; private land ownership allowed.

1991 March: Yeltsin secured the support of Congress of Peoples' Deputies for direct election of an executive president. June: Yeltsin elected president under a liberal-radical banner. Aug: Yeltsin stood out against abortive anti-Gorbachev coup, emerging as key power-broker within Soviet Union; pre-revolutionary flag restored. Sept: Silaev resigned as Russian premier. Nov: Yeltsin also named as prime minister; Communist Party of the Soviet Union and Russian Communist Party banned; Yeltsin's goverment gained control of Russia's economic assets and armed forces. Dec: Yeltsin negotiated formation of new confederal Commonwealth of Independent States (CIS); Russia admitted into United Nations (UN); independence recognized by USA and European Community (EC).

1992 Jan: admitted into Conference on Security and Cooperation in Europe (CSCE); assumed former USSR's permanent seat on UN Security Council; prices freed. Feb: demonstrations in Moscow and other cities as living standards plummeted. March: 18 out of 20 republics signed treaty agreeing to remain within loose Russian Federation; Tatarstan and Checheno-Ingush refused to sign. Aug: agreement with Ukraine on joint control of Black Sea fleet. Dec: Victor Chernomyrdin elected prime minister; new constitution agreed in referendum. START II arms-reduction agreement signed with USA.

1993 March: Congress of People's Deputies attempted to limit Yeltsin's powers to rule by decree and cancel referendum due 25 April. Yeltsin declared temporary presidential 'special rule' pending holding of referendum. Results of referendum showed Russians in favour of Yeltsin's reforms.

Rwanda
 Republic of
(*Republika y'u
Rwanda*)

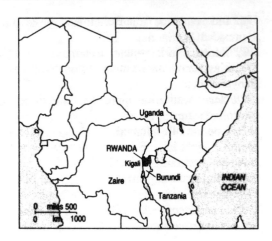

area 26,338 sq km/10,173 sq mi
capital Kigali
towns Butare, Ruhengeri
physical high savanna and hills, with volcanic mountains in NW
head of state and government Maj-Gen Juvenal Habyarimana
from 1973
political system one-party military republic
exports coffee, tea, pyrethrum
currency franc
population (1990 est) 7,603,000 (Hutu 90%, Tutsi 9%, Twa 1%);
growth rate 3.3% p.a.
life expectancy men 49, women 53 (1989)
languages Kinyarwanda, French (official); Kiswahili
religions Roman Catholic 54%, animist 23%, Protestant 12%;
Muslim 9%
literacy men 50% (1989)
GNP $2.3 bn (1987); $323 per head (1986)
chronology
1916 Belgian troops occupied Rwanda; League of Nations mandated
Rwanda and Burundi to Belgium as Territory of Ruanda-Urundi.
1959 Interethnic warfare between Hutu and Tutsi.

1962 Independence from Belgium achieved, with Grégoire Kayibanda as president.

1972 Renewal of interethnic fighting.

1973 Kayibanda ousted in a military coup led by Maj-Gen Juvenal Habyarimana.

1978 New constitution approved; Rwanda remained a military-controlled state.

1980 Civilian rule adopted.

1988 Refugees from Burundi massacres streamed into Rwanda.

1990 Government attacked by Rwandan Patriotic Front (FPR), a Tutsi military-political organization based in Uganda. Constitutional reforms promised.

1992 Peace accord with FPR.

1993 Power-sharing agreement with government repudiated by FPR.

St Christopher (St Kitts)–Nevis
Federation of

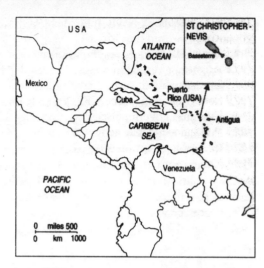

area 269 sq km/104 sq mi (St Christopher 176 sq km/68 sq mi, Nevis 93 sq km/36 sq mi)

capital Basseterre (on St Christopher)

towns Charlestown (largest on Nevis)

physical both islands are volcanic

head of state Elizabeth II from 1983, represented by governor general

head of government Kennedy Simmonds from 1980

political system federal constitutional monarchy

exports sugar, molasses, electronics, clothing

currency Eastern Caribbean dollar

population (1990 est) 45,800; growth rate 0.2% p.a.

life expectancy men 69, women 72

language English

religion Anglican 36%, Methodist 32%, other Protestant 8%, Roman Catholic 10% (1985 est)

literacy 90% (1987)

GNP $40 million (1983); $870 per head

chronology

1871–1956 Part of the Leeward Islands Federation.

1958–62 Part of the Federation of the West Indies.

1967 St Christopher, Nevis, and Anguilla achieved internal self-government, within the British Commonwealth, with Robert Bradshaw, Labour Party leader, as prime minister.

1971 Anguilla returned to being a British dependency.

1978 Bradshaw died; succeeded by Paul Southwell.

1979 Southwell died; succeeded by Lee L Moore.

1980 Coalition government led by Kennedy Simmonds.

1983 Full independence achieved within the Commonwealth.

1984 Coalition government re-elected.

1989 Prime Minister Simmonds won a third successive term.

St Lucia

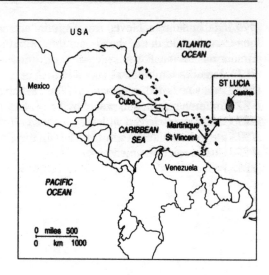

area 617 sq km/238 sq mi

capital Castries

towns Vieux-Fort, Soufrière

physical mountainous island with fertile valleys; mainly tropical forest

head of state Elizabeth II from 1979, represented by governor general

head of government John Compton from 1982

political system constitutional monarchy

exports coconut oil, bananas, cocoa, copra

currency Eastern Caribbean dollar

population (1990 est) 153,000; growth rate 2.8% p.a.

life expectancy men 68, women 73 (1989)

languages English; French patois

religion Roman Catholic 90%

literacy 78% (1989)

GNP $166 million; $1,370 per head (1987)

chronology

1814 Became a British crown colony following Treaty of Paris.

1967 Acquired internal self-government as a West Indies associated state.

1979 Independence achieved from Britain, within the Commonwealth.
John Compton, leader of the United Workers' Party (UWP), became
prime minister. Allan Louisy, leader of the St Lucia Labour Party
(SLP), replaced Compton as prime minister.
1981 Louisy resigned; replaced by Winston Cenac.
1982 Compton returned to power at the head of a UWP government.
1987 Compton re-elected with reduced majority.
1991 Integration with Windward Islands proposed.
1992 UWP won general election.

St Vincent and the Grenadines

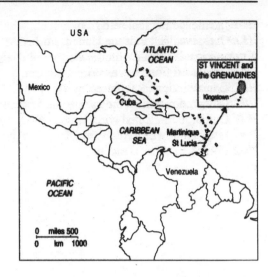

area 388 sq km/150 sq mi, including islets of the Northern Grenadines 43 sq km/17 sq mi
capital Kingstown
towns Georgetown, Chateaubelair
physical volcanic mountains, thickly forested
head of state Elizabeth II from 1979, represented by governor general
head of government James Mitchell from 1984
political system constitutional monarchy
exports bananas, taros, sweet potatoes, arrowroot, copra
currency Eastern Caribbean dollar
population (1990 est) 106,000; growth rate –4% p.a.
life expectancy men 69, women 74 (1989)
languages English; French patois
religions Anglican 47%, Methodist 28%, Roman Catholic 13%
literacy 85% (1989)
GNP $188 million; $1,070 per head (1987)
chronology
1783 Became a British crown colony.
1958–62 Part of the West Indies Federation.

1969 Achieved internal self-government.
1979 Achieved full independence from Britain, within the Commonwealth, with Milton Cato as prime minister.
1984 James Mitchell replaced Cato as prime minister.
1989 Mitchell decisively re-elected.
1991 Integration with Windward Islands proposed.

Samoa, Western
Independent State of
(*Samoa i Sisifo*)

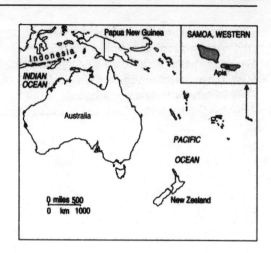

area 2,830 sq km/1,093 sq mi
capital Apia (on Upolu island)
physical comprises South Pacific islands of Savai'i and Upolu, with two smaller tropical islands and islets; mountain ranges on main islands
head of state King Malietoa Tanumafili II from 1962
head of government Tofilau Eti Alesana from 1988
political system liberal democracy
exports coconut oil, copra, cocoa, fruit juice, cigarettes, timber
currency talà
population (1989) 169,000; growth rate 1.1% p.a.
life expectancy men 64, women 69 (1989)
languages English, Samoan (official)
religions Protestant 70%, Roman Catholic 20%
literacy 90% (1989)
GNP $110 million (1987); $520 per head
chronology
1899–1914 German protectorate.
1920–61 Administered by New Zealand.
1959 Local government elected.
1961 Referendum favoured independence.

1962 Independence achieved within the Commonwealth, with Fiame
Mata Afa Mulinu'u as prime minister.
1975 Mata Afa died.
1976 Tupuola Taisi Efi became first nonroyal prime minister.
1982 Va'ai Kolone became prime minister; replaced by Tupuola Efi.
Assembly failed to approve budget; Tupuola Efi resigned; replaced by
Tofilau Eti Alesana.
1985 Tofilau Eti resigned; head of state invited Va'ai Kolone to lead
the government.
1988 Elections produced a hung parliament, with first Tupuola Efi as
prime minister and then Tofilau Eti Alesana.
1990 Universal adult suffrage introduced.
1991 Tofilau Eti Alesana re-elected. Fiame Naome became first
woman in cabinet.

San Marino
Republic of
(*Repubblica di San Marino*)

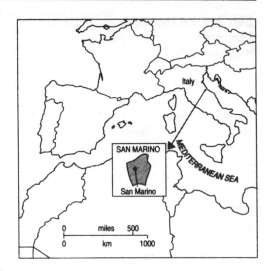

area 61 sq km/24 sq mi
capital San Marino
towns Serravalle (industrial centre)
physical on the slope of Mount Titano
heads of state and government two captains regent, elected for a six-month period
political system direct democracy
exports wine, ceramics, paint, chemicals, building stone
currency Italian lira
population (1990 est) 23,000; growth rate 0.1% p.a.
life expectancy men 70, women 77
language Italian
religion Roman Catholic 95%
literacy 97% (1987)
chronology
1862 Independence recognized under Italy's protection.
1947–86 Governed by a series of left-wing and centre-left coalitions.
1986 Communist and Christian Democrat 'grand coalition'.
1992 Joined the United Nations.

São Tomé e Príncipe
Democratic
Republic of

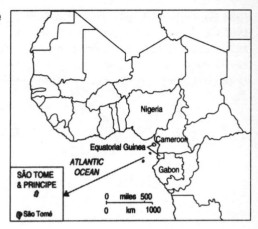

area 1,000 sq km/386 sq mi
capital São Tomé
towns Santo Antonio, Santa Cruz
physical comprises two main islands and several smaller ones, all volcanic; thickly forested and fertile
head of state and government Miguel Trovoada from 1991
political system emergent democratic republic
exports cocoa, copra, coffee, palm oil and kernels
currency dobra
population (1990 est) 125,000; growth rate 2.5% p.a.
life expectancy men 62, women 62
languages Portuguese (official), Fang (Bantu)
religions Roman Catholic 80%, animist
literacy men 73%, women 42% (1981)
GNP $32 million (1987); $384 per head (1986)
chronology
1471 Discovered by Portuguese.
1522–1973 A province of Portugal.
1973 Achieved internal self-government.
1975 Independence achieved from Portugal, with Manuel Pinto da Costa as president.

1984 Formally declared a nonaligned state.
1987 Constitution amended.
1988 Unsuccessful coup attempt against da Costa.
1990 New constitution approved.
1991 First multiparty elections held; Miguel Trovoada replaced Pinto da Costa.

Saudi Arabia
Kingdom of
(*al-Mamlaka al-
'Arabiya as-
Sa'udiya*)

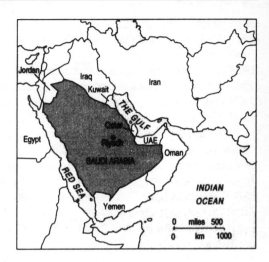

area 2,200,518 sq km/849,400 sq mi
capital Riyadh
towns Mecca, Medina, Taif, Jidda, Dammam
physical desert, sloping to the Persian Gulf from a height of 2,750 m/
9,000 ft in the W
environment oil pollution caused by the Gulf War 1990–91 has
affected 460 km/285 mi of the Saudi coastline, threatening
desalination plants and damaging the wildlife of saltmarshes,
mangrove forest, and mudflats
head of state and government King Fahd Ibn Abdul Aziz from 1982
political system absolute monarchy
exports oil, petroleum products
currency rial
population (1990 est) 16,758,000 (16% nomadic); growth rate
3.1% p.a.
life expectancy men 64, women 67 (1989)
language Arabic
religion Sunni Muslim; there is a Shi'ite minority
literacy men 34%, women 12% (1980 est)
GNP $70 bn (1988); $6,170 per head (1988)

chronology
1926–32 Territories of Nejd and Hejaz united and kingdom established.
1953 King Ibn Saud died and was succeeded by his eldest son, Saud.
1964 King Saud forced to abdicate; succeeded by his brother, Faisal.
1975 King Faisal assassinated; succeeded by his half-brother, Khalid.
1982 King Khalid died; succeeded by his brother, Crown Prince Fahd.
1987 Rioting by Iranian pilgrims caused 400 deaths in Mecca; diplomatic relations with Iran severed.
1990 Iraqi troops invaded and annexed Kuwait and massed on Saudi Arabian border. King Fahd called for help from US and UK forces.
1991 King Fahd provided military and financial assistance in Gulf War. Calls from religious leaders for 'consultative assembly' to assist in government of kingdom. Saudi Arabia attended Middle East peace conference.
1992 Formation of a 'consultative council' seen as possible move towards representative government.

Senegal
Republic of
(*République du
Sénégal*)

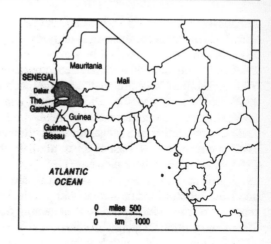

area 196,200 sq km/75,753 sq mi
capital (and chief port) Dakar
towns Thiès, Kaolack
physical plains rising to hills in SE; swamp and tropical forest in SW
head of state and government Abdou Diouf from 1981
political system emergent socialist democratic republic
exports peanuts, cotton, fish, phosphates
currency franc CFA
population (1990 est) 7,740,000; growth rate 3.1% p.a.
life expectancy men 51, women 54 (1989)
languages French (official); African dialects are spoken
religions Muslim 80%, Roman Catholic 10%, animist
literacy men 37%, women 19% (1985 est)
GNP $2 bn (1987); $380 per head (1984)
chronology
1659 Became a French colony.
1854–65 Interior occupied by French.
1902 Became a territory of French West Africa.
1959 Formed the Federation of Mali with French Sudan.
1960 Independence achieved from France, but withdrew from the
federation. Léopold Sédar Senghor, leader of the Senegalese

Progressive Union (UPS), became president.
1966 UPS declared the only legal party.
1974 Pluralist system re-established.
1976 UPS reconstituted as Senegalese Socialist Party (PS). Prime Minister Abdou Diouf nominated as Senghor's successor.
1980 Senghor succeeded by Diouf. Troops sent to defend Gambia.
1981 Military help again sent to Gambia.
1982 Confederation of Senegambia came into effect.
1983 Diouf re-elected. Post of prime minister abolished.
1988 Diouf decisively re-elected.
1989 Violent clashes between Senegalese and Mauritanians in Dakar and Nouakchott killed more than 450 people; over 50,000 people repatriated from both countries. Senegambia federation abandoned.
1991 Constitutional changes outlined.
1992 Diplomatic links with Mauritania re-established.

Seychelles
Republic of

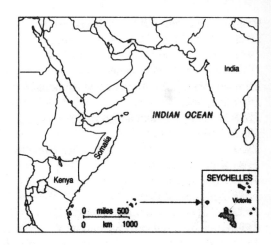

area 453 sq km/175 sq mi
capital Victoria (on Mahé island)
towns Cascade, Port Glaud, Misere
physical comprises two distinct island groups, one concentrated, the other widely scattered, totalling over 100 islands and islets
head of state and government France-Albert René from 1977
political system one-party socialist republic
exports copra, cinnamon
currency Seychelles rupee
population (1990) 71,000; growth rate 2.2% p.a.
life expectancy 66 years (1988)
languages creole (Asian, African, European mixture) 95%, English, French (all official)
religion Roman Catholic 90%
literacy 80% (1989)
GNP $175 million; $2,600 per head (1987)
chronology
1744 Became a French colony.
1794 Captured by British.
1814 Ceded by France to Britain; incorporated as a dependency of Mauritius.

1903 Became a separate British colony.

1975 Internal self-government agreed.

1976 Independence achieved from Britain as a republic within the Commonwealth, with James Mancham as president.

1977 France-Albert René ousted Mancham in an armed coup and took over presidency.

1979 New constitution adopted; Seychelles People's Progressive Front (SPPF) sole legal party.

1981 Attempted coup by South African mercenaries thwarted.

1984 René re-elected.

1987 Coup attempt foiled.

1989 René re-elected.

1991 Multiparty politics promised.

1992 Mancham returned from exile. Constitutional commission elected; referendum on constitutional reform received insufficient support.

Sierra Leone
Republic of

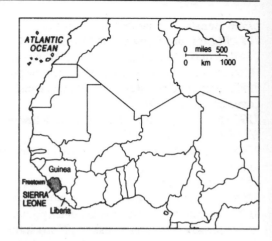

area 71,740 sq km/27,710 sq mi
capital Freetown
towns Koidu, Bo, Kenema, Makeni
physical mountains in E; hills and forest; coastal mangrove swamps
head of state and government military council headed by Capt
Valentine Strasser from 1992
political system transitional
exports palm kernels, cocoa, coffee, ginger, diamonds, bauxite, rutile
currency leone
population (1990 est) 4,168,000; growth rate 2.5% p.a.
life expectancy men 41, women 47 (1989)
languages English (official), local languages
religions animist 52%, Muslim 39%, Protestant 6%, Roman Catholic
2% (1980 est)
literacy men 38%, women 21% (1985 est)
GNP $965 million (1987); $320 per head (1984)
chronology
1808 Became a British colony.
1896 Hinterland declared a British protectorate.
1961 Independence achieved from Britain within the Commonwealth,
with Milton Margai, leader of Sierra Leone People's Party (SLPP), as

prime minister.

1964 Milton succeeded by his half-brother, Albert Margai.

1967 Election results disputed by army, who set up a National Reformation Council and forced the governor general to leave.

1968 Army revolt made Siaka Stevens, leader of the All People's Congress (APC), prime minister.

1971 New constitution adopted, making Sierra Leone a republic, with Stevens as president.

1978 APC declared only legal party. Stevens sworn in for another seven-year term.

1985 Stevens retired; succeeded by Maj-Gen Joseph Momoh.

1989 Attempted coup against President Momoh foiled.

1991 Referendum endorsed multiparty politics.

1992 Military take-over; President Momoh fled. National Provisional Ruling Council (NPRC) established under Capt Valentine Strasser.

Singapore
Republic of

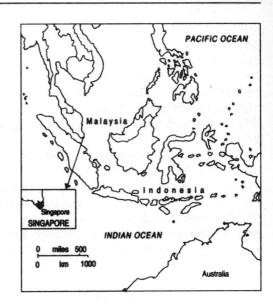

area 622 sq km/240 sq mi
capital Singapore City
towns Jurong, Changi
physical comprises Singapore Island, low and flat, and 57 small islands
head of state Wee Kim Wee from 1985
head of government Goh Chok Tong from 1990
political system liberal democracy with strict limits on dissent
exports electronics, petroleum products, rubber, machinery, vehicles
currency Singapore dollar
population (1990 est) 2,703,000 (Chinese 75%, Malay 14%, Tamil 7%); growth rate 1.2% p.a.
life expectancy men 71, women 77 (1989)
languages Malay (national tongue), Chinese, Tamil, English (all official)
religions Buddhist, Taoist, Muslim, Hindu, Christian
literacy men 93%, women 79% (1985 est)

GDP $19.9 bn (1987); $7,616 per head
chronology
1819 Singapore leased to British East India Company.
1858 Placed under rule of British crown.
1942 Invaded and occupied by Japan.
1945 Japanese removed by British forces.
1959 Independence achieved from Britain; Lee Kuan Yew became prime minister.
1963 Joined new Federation of Malaysia.
1965 Left federation to become an independent republic.
1984 Opposition made advances in parliamentary elections.
1986 Opposition leader convicted of perjury and prohibited from standing for election.
1988 Ruling conservative party elected to all but one of available assembly seats; increasingly authoritarian rule.
1990 Lee Kuan Yew resigned as prime minister; replaced by Goh Chok Tong.
1991 People's Action Party (PAP) and Goh Chok Tong re-elected.

Slovak Republic
(Slovenská Republika)

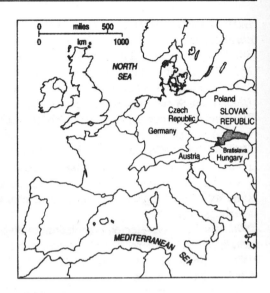

area 49,035 sq km/18,940 sq mi
capital Bratislava
towns Kosice, Nitra, Presov, Banská Bystrica
physical Carpathian Mountains including Tatra and Beskids in N; fine beech and oak forests; Danube plain in S
head of state Michal Kovak from 1993
head of government Vladimir Meciar from 1993
political system emergent democracy
exports iron ore, copper, mercury, magnesite, armaments, chemicals, textiles, machinery
currency new currency based on koruna
population (1991) 5,268,900 (with Hungarian and other minorities); growth rate 0.4% p.a.
life expectancy men 68, women 75
languages Slovak (official)
religions Roman Catholic (over 50%), Lutheran, Reformist, Orthodox
literacy 100%
GDP $10,000 million (1990); $1,887 per head

chronology

906–1918 Under Magyar domination.

1918 Independence achieved from Austro-Hungarian Empire; Slovaks joined Czechs in forming Czechoslovakia as independent nation.

1948 Communists assumed power in Czechoslovakia.

1968 Slovak Socialist Republic created under new federal constitution.

1989 Pro-democracy demonstrations in Prague and Bratislava; new political parties formed, including Slovak-based People Against Violence (PAV); Communist Party stripped of powers. Dec: new 'grand coalition' government formed, including former dissidents; political parties legalized; Václav Havel appointed state president. Amnesty granted to 22,000 prisoners; calls for USSR to withdraw its troops.

1990 July: Havel re-elected president in multiparty elections.

1991 Evidence of increasing Czech and Slovak separatism. March: PAV splinter group formed under Slovak premier Vladimir Meciar. April: Meciar dismissed, replaced by Jan Carnogursky; pro-Meciar rallies held in Bratislava. July: Soviet troops withdrawn. Oct: Public Against Violence renamed Civic Democratic Union–Public Against Violence (PAV).

1992 March: PAV renamed Civic Democratic Union (CDU). June: Havel resigned following Slovak gains in assembly elections. Aug: agreement on creation of separate Czech and Slovak states from Jan 1993.

1993 Jan: Slovak Republic became sovereign state, with Meciar, leader of the Movement for a Democratic Slovakia (MFDS), as prime minister. Feb: Michal Kovak became president.

Slovenia
Republic of

area 20,251 sq km/7,817 sq mi
capital Ljubljana
towns Maribor, Kranj, Celji, Koper
physical mountainous; Sava and Drava rivers
head of state Milan Kucan from 1990
head of government Janez Drnovsek from 1992
political system emergent democracy
products grain, sugarbeet, livestock, timber, cotton and woollen
textiles, steel, vehicles
currency tolar
population (1990) 2,000,000 (Slovene 91%, Croat 3%, Serb 2%)
languages Slovene, resembling Serbo-Croat
religion Roman Catholic
chronology
1918 United with Serbia and Croatia.
1929 The kingdom of Serbs, Croats, and Slovenes took the name of
Yugoslavia.
1945 Became a constituent republic of Yugoslav Socialist Federal
Republic.

mid-1980s The Slovenian Communist Party liberalized itself and agreed to free elections. Yugoslav counterintelligence (KOV) began repression.

1989 Jan: Social Democratic Alliance of Slovenia launched as the first political organization independent of Communist Party.
Sept: constitution changed to allow secession from federation.

1990 Feb: Slovene League of Communists, renamed as the Party of Democratic Reform, severed its links with the Yugoslav League of Communists. April: nationalist Democratic Opposition of Slovenia (DEMOS) coalition secured victory in first multiparty parliamentary elections; Milan Kucan became president. July: sovereignty declared.
Dec: independence overwhelmingly approved in referendum.

1991 June: independence declared; 100 killed after federal army intervened; cease-fire brokered by European Community (EC).
July: cease-fire agreed between federal troops and nationalists. Oct: withdrawal of Yugoslav army completed. Dec: DEMOS coalition dissolved.

1992 Jan: EC recognized Slovenia's independence. April: Janez Drnovsek appointed prime minister designate; independence recognized by USA. May: admitted into United Nations and Conference on Security and Cooperation in Europe (CSCE).
Dec: Liberal Democrats and Christian Democrats won assembly elections; Kucan re-elected president.

1993 Drnovsek re-elected prime minister.

Solomon Islands

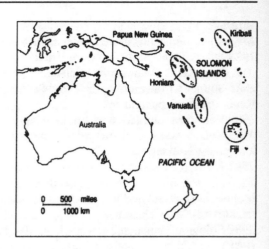

area 27,600 sq km/10,656 sq mi
capital Honiara (on Guadalcanal)
towns Gizo, Yandina
physical comprises all but the northernmost islands (which belong to Papua New Guinea) of a Melanesian archipelago stretching nearly 1,500 km/900 mi. The largest is Guadalcanal (area 6,500 sq km/2,510 sq mi); others are Malaita, San Cristobal, New Georgia, Santa Isabel, Choiseul; mainly mountainous and forested
head of state Elizabeth II represented by governor general
head of government Solomon Mamaloni from 1989
political system constitutional monarchy
exports fish products, palm oil, copra, cocoa, timber
currency Solomon Island dollar
population (1990 est) 314,000 (Melanesian 95%, Polynesian 4%); growth rate 3.9% p.a.
life expectancy men 66, women 71
languages English (official); there are some 120 Melanesian dialects
religions Anglican 34%, Roman Catholic 19%, South Sea Evangelical 17%
literacy 60% (1989)
GNP $141 million; $420 per head (1987)

chronology

1893 Solomon Islands placed under British protection.

1978 Independence achieved from Britain within the Commonwealth, with Peter Kenilorea as prime minister.

1981 Solomon Mamaloni of the People's Progressive Party replaced Kenilorea as prime minister.

1984 Kenilorea returned to power, heading a coalition government.

1986 Kenilorea resigned after allegations of corruption; replaced by his deputy, Ezekiel Alebua.

1988 Kenilorea elected deputy prime minister. Joined Vanuatu and Papua New Guinea to form the Spearhead Group, aiming to preserve Melanesian cultural traditions and secure independence for the French territory of New Caledonia.

1989 Solomon Mamaloni, now leader of the People's Alliance Party (PAP), elected prime minister; formed PAP-dominated coalition.

1990 Mamaloni resigned as PAP party leader, but continued as head of a government of national unity.

Somalia
Somali Democratic
Republic
(*Jamhuriyadda
Dimugradiga
Somaliya*)

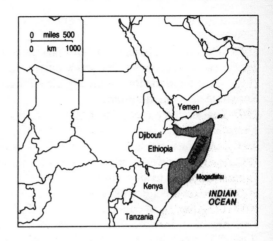

area 637,700 sq km/246,220 sq mi
capital Mogadishu
towns Hargeisa, Kismayu, Berbera
physical mainly flat, with hills in N
environment destruction of trees for fuel and by grazing livestock has
led to an increase in desert area
head of state and government Ali Mahdi Mohammed from 1991
political system one-party socialist republic
exports livestock, skins, hides, bananas, fruit
currency Somali shilling
population (1990 est) 8,415,000 (including 350,000 refugees in
Ethiopia and 50,000 in Djibouti); growth rate 3.1% p.a.
life expectancy men 53, women 53 (1989)
languages Somali, Arabic (both official), Italian, English
religion Sunni Muslim 99%
literacy 40% (1986)
GNP $1.5 bn; $290 per head (1987)
chronology
1884–87 British protectorate of Somaliland established.
1889 Italian protectorate of Somalia established.
1960 Independence achieved from Italy and Britain.

1963 Border dispute with Kenya; diplomatic relations broken with Britain.

1968 Diplomatic relations with Britain restored.

1969 Army coup led by Maj-Gen Mohamed Siad Barre; constitution suspended, Supreme Revolutionary Council set up; name changed to Somali Democratic Republic.

1978 Defeated in eight-month war with Ethiopia. Armed insurrection began in north.

1979 New constitution for socialist one-party state adopted.

1982 Antigovernment Somali National Movement formed. Oppressive countermeasures by government.

1987 Barre re-elected president.

1989 Dissatisfaction with government and increased guerrilla activity in north.

1990 Civil war intensified. Constitutional reforms promised.

1991 Mogadishu captured by rebels; Barre fled; Ali Mahdi Mohammed named president; free elections promised. Secession of NE Somalia, as the Somaliland Republic, announced. Cease-fire signed, but later collapsed. Thousands of casualties as a result of heavy fighting in capital.

1992 Relief efforts to ward off impending famine severely hindered by unstable political situation; relief convoys hijacked by 'war lords'. Dec: UN peacekeeping troops, mainly US Marines, drafted in to protect relief operations; dominant warlords agreed truce.

1993 March: leaders of armed factions agreed to federal system of government, based on 18 autonomous regions.

South Africa
Republic of
(*Republiek van
Suid-Afrika*)

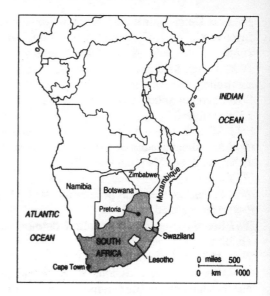

area 1,223,181 sq km/472,148 sq mi (includes Walvis Bay and independent black homelands)
capital Cape Town (legislative), Pretoria (administrative),
Bloemfontein (judicial)
towns Johannesburg, Durban, Port Elizabeth, East London
physical southern end of large plateau, fringed by mountains and lowland coastal margin
territories Marion Island and Prince Edward Island in the Antarctic
head of state and government F W de Klerk from 1989
political system nationalist republic, restricted democracy
exports maize, sugar, fruit, wool, gold (world's largest producer), platinum, diamonds, uranium, iron and steel, copper; mining and minerals are largest export industry, followed by arms manufacturing
currency rand
population (1990 est) 39,550,000 (73% black: Zulu, Xhosa, Sotho, Tswana; 18% white: 3% mixed, 3% Asian); growth rate 2.5% p.a.
life expectancy whites 71, Asians 67, blacks 58
languages Afrikaans and English (both official), Bantu

religions Dutch Reformed Church 40%, Anglican 11%, Roman
Catholic 8%, other Christian 25%, Hindu, Muslim
literacy whites 99%, Asians 69%, blacks 50% (1989)
GNP $81 bn; $1,890 per head (1987)
chronology
1910 Union of South Africa formed from two British colonies and two
Boer republics.
1912 African National Congress (ANC) formed.
1948 Apartheid system of racial discrimination initiated by Daniel
Malan, leader of National Party (NP).
1955 Freedom Charter adopted by ANC.
1958 Malan succeeded as prime minister by Hendrik Verwoerd.
1960 ANC banned.
1961 South Africa withdrew from Commonwealth and became a
republic.
1962 ANC leader Nelson Mandela jailed.
1964 Mandela, Walter Sisulu, Govan Mbeki, and five other ANC
leaders sentenced to life imprisonment.
1966 Verwoerd assassinated; succeeded by B J Vorster.
1976 Soweto uprising.
1977 Death in custody of Pan African Congress (PAC) activist
Steve Biko.
1978 Vorster resigned and was replaced by P W Botha.
1984 New constitution adopted, giving segregated representation to
Coloureds and Asians and making Botha president. Nonaggression
pact with Mozambique signed but not observed.
1985 Growth of violence in black townships.
1986 Commonwealth agreed on limited sanctions. US Congress voted
to impose sanctions. Some major multinational companies closed
down their South African operations.
1987 Government formally acknowledged the presence of its military
forces in Angola.
1988 Botha announced 'limited constitutional reforms'. South Africa
agreed to withdraw from Angola and recognize Namibia's
independence as part of regional peace accord.
1989 Botha gave up NP leadership and state presidency. F W de Klerk

became president. ANC activists released; beaches and public facilities desegregated. Elections held in Namibia to create independence government.

1990 ANC ban lifted; Nelson Mandela released from prison. NP membership opened to all races. ANC leader Oliver Tambo returned. Daily average of 35 murders and homicides recorded.

1991 Mandela and Zulu leader Chief Gatsha Buthelezi urged end to fighting between ANC and Inkatha. Mandela elected ANC president. Revelations of government support for Inkatha threatened ANC cooperation. De Klerk announced repeal of remaining apartheid laws. South Africa readmitted to international sport. USA lifted sanctions. PAC and Buthelezi withdrew from negotiations over new constitution.

1992 Constitution leading to all-races majority rule approved by whites-only referendum. Massacre of civilians at black township of Boipathong near Johannesburg by Inkatha, aided and abetted by police, threatened constitutional talks.

1993 Feb: de Klerk and Nelson Mandela agreed to formation of government of national unity after free elections late 1993/early 1994. Buthelezi not consulted; he opposed such an arrangement. April: ANC leader Chris Hani assassinated; Andries Treurnicht, leader of Conservative Party, and Oliver Tambo died.

Spain
(*España*)

area 504,750 sq km/194,960 sq mi
capital Madrid
towns Seville, Barcelona, Valencia, Málaga, Cádiz
physical central plateau with mountain ranges; lowlands in S
territories Balearic and Canary Islands; in N Africa: Ceuta, Melilla,
Alhucemas, Chafarinas Is, Peñón de Vélez
head of state King Juan Carlos I from 1975
head of government Felipe González Márquez from 1982
political system constitutional monarchy
exports citrus fruits, grapes, pomegranates, vegetables, wine, sherry,
olive oil, canned fruit and fish, iron ore, cork, vehicles, textiles,
petroleum products, leather goods, ceramics
currency peseta
population (1990 est) 39,623,000; growth rate 0.2% p.a.
life expectancy men 74, women 80 (1989)
languages Spanish (Castilian, official), Basque, Catalan, Galician,
Valencian, Majorcan
religion Roman Catholic 99%
literacy 97% (1989)

GNP $288 bn (1987); $4,490 per head (1984)
recent chronology
1936–39 Civil war; General Francisco Franco became head of state and government; fascist party Falange declared only legal political organization.
1947 General Franco announced restoration of the monarchy after his death, with Prince Juan Carlos as his successor.
1975 Franco died; succeeded as head of state by King Juan Carlos I.
1978 New constitution adopted with Adolfo Suárez, leader of the Democratic Centre Party, as prime minister.
1981 Suárez resigned; succeeded by Leopoldo Calvo Sotelo. Attempted military coup thwarted.
1982 Socialist Workers' Party (PSOE), led by Felipe González, won a sweeping electoral victory. Basque separatist organization (ETA) stepped up its guerrilla campaign.
1985 ETA's campaign spread to holiday resorts.
1986 Referendum confirmed NATO membership. Spain joined the European Economic Community.
1988 Spain joined the Western European Union.
1989 PSOE lost seats to hold only parity after general election. Talks between government and ETA collapsed and truce ended.
1992 ETA's 'armed struggle' resumed. Nov: Maastricht Treaty ratified by parliament.

Sri Lanka

Democratic
Socialist Republic
of (*Prajathanrika
Samajawadi
Janarajaya Sri
Lanka*)
(until 1972 *Ceylon*)

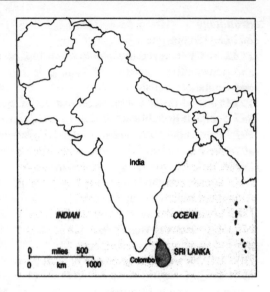

area 65,600 sq km/25,328 sq mi
capital (and chief port) Colombo
towns Kandy, Jaffna, Galle, Negombo, Trincomalee
physical flat in N and around the coast; hills and mountains in S and central interior
head of state Ranasinghe Premadasa from 1989
head of government Dingiri Banda Wijetunge from 1989
political system liberal democratic republic
exports tea, rubber, coconut products, graphite, sapphires, rubies, other gemstones
currency Sri Lanka rupee
population (1990 est) 17,135,000 (Sinhalese 74%, Tamils 17%, Moors 7%); growth rate 1.8% p.a.
life expectancy men 67, women 72 (1989)
languages Sinhala, Tamil, English
religions Buddhist 69%, Hindu 15%, Muslim 8%, Christian 7%
literacy 87% (1988)
GNP $7.2 bn; $400 per head (1988)

chronology
1802 Ceylon became a British colony.
1948 Ceylon achieved independence from Britain within the Commonwealth.
1956 Sinhala established as the official language.
1959 Prime Minister Solomon Bandaranaike assassinated.
1972 Socialist Republic of Sri Lanka proclaimed.
1978 Presidential constitution adopted by new government led by Junius Jayawardene of the United National Party (UNP).
1983 Tamil guerrilla violence escalated; state of emergency imposed.
1987 President Jayawardene and Indian prime minister Rajiv Gandhi signed Colombo Accord. Violence continued despite cease-fire policed by Indian troops.
1988 Left-wing guerrillas campaigned against Indo-Sri Lankan peace pact. Prime Minister Ranasinghe Premadasa elected president.
1989 Premadasa became president; D B Wijetunge, prime minister. Leaders of the Tamil United Liberation Front (TULF) and terrorist People's Liberation Front assassinated.
1990 Indian peacekeeping force withdrawn. Violence continued.
1991 March: defence minister, Ranjan Wijeratne, assassinated; Sri Lankan army killed 2,552 Tamil Tigers at Elephant Pass. October: impeachment motion against President Premadasa failed. Dec: new party, the Democratic United National Front, formed by former members of UNP.

Sudan
Democratic
Republic of
(*Jamhuryat
es-Sudan*)

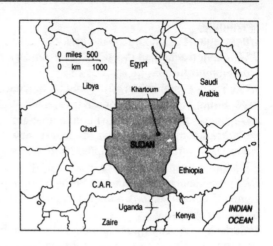

area 2,505,800 sq km/967,489 sq mi
capital Khartoum
towns Omdurman, Juba, Wadi Medani, Kassala, Atbara, Kosti,
Port Sudan
physical fertile valley of river Nile separates Libyan Desert in W from
high rocky Nubian Desert in E
environment the building of the Jonglei Canal to supply water to
N Sudan and Egypt threatens the grasslands of S Sudan
head of state and government General Omar Hassan Ahmed el-Bashir
from 1989
political system military republic
exports cotton, gum arabic, sesame seed, peanuts, sorghum
currency Sudanese pound
population (1990 est) 25,164,000; growth rate 2.9% p.a.
life expectancy men 51, women 55 (1989)
languages Arabic 51% (official), local languages
religions Sunni Muslim 73%, animist 18%, Christian 9% (in south)
literacy 30% (1986)
GNP $8.5 bn (1988); $330 per head (1988)
recent chronology
1956 Sudan achieved independence from Britain and Egypt as

a republic.

1958 Military coup replaced civilian government with Supreme Council of the Armed Forces.

1964 Civilian rule reinstated.

1969 Coup led by Col Gaafar Mohammed Nimeri established Revolutionary Command Council (RCC); name changed to Democratic Republic of Sudan.

1970 Union with Egypt agreed in principle.

1971 New constitution adopted; Nimeri confirmed as president; Sudanese Socialist Union (SSU) declared only legal party.

1972 Proposed Federation of Arab Republics, comprising Sudan, Egypt, and Syria, abandoned. Addis Ababa conference proposed autonomy for southern provinces.

1974 National assembly established.

1983 Nimeri re-elected. Shari'a (Islamic law) introduced.

1985 Nimeri deposed in a bloodless coup led by General Swar al-Dahab; transitional military council set up. State of emergency declared.

1986 More than 40 political parties fought general election; coalition government formed.

1987 Virtual civil war with Sudan People's Liberation Army (SPLA).

1988 Al-Mahdi formed a new coalition. Another flare-up of civil war between north and south created tens of thousands of refugees. Floods made 1.5 million people homeless. Peace pact signed with SPLA.

1989 Sadiq al-Mahdi overthrown in coup led by General Omar Hassan Ahmed el-Bashir.

1990 Civil war continued with new SPLA offensive.

1991 Federal system introduced, with division of country into nine states.

1993 March: SPLA leaders John Garang and Riek Machar announced unilateral cease-fire in ten years' war with government in Khartoum.

Surinam
Republic of
(*Republiek
Suriname*)

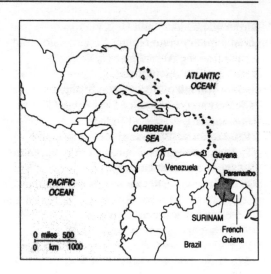

area 163,820 sq km/63,243 sq mi
capital Paramaribo
towns Nieuw Nickerie, Brokopondo, Nieuw Amsterdam
physical hilly and forested, with flat and narrow coastal plain
head of state and government Ronald Venetiaan from 1991
political system emergent democratic republic
exports alumina, aluminium, bauxite, rice, timber
currency Surinam guilder
population (1990 est) 408,000 (Hindu 37%, Creole 31%, Javanese 15%); growth rate 1.1% p.a.
life expectancy men 66, women 71 (1989)
languages Dutch (official), Sranan (creole), English, others
religions Christian 30%, Hindu 27%, Muslim 20%
literacy 65% (1989)
GNP $1.1 bn (1987); $2,920 per head (1985)
chronology
1667 Became a Dutch colony.
1954 Achieved internal self-government as Dutch Guiana.
1975 Independence achieved from the Netherlands, with Dr Johan

Ferrier as president and Henck Arron as prime minister; 40% of the population emigrated to the Netherlands.

1980 Arron's government overthrown in army coup; Ferrier refused to recognize military regime; appointed Dr Henk Chin A Sen to lead civilian administration. Army replaced Ferrier with Dr Chin A Sen.

1982 Army, led by Lt Col Desi Bouterse, seized power, setting up a Revolutionary People's Front.

1985 Ban on political activities lifted.

1986 Antigovernment rebels brought economic chaos to Surinam.

1987 New constitution approved.

1988 Ramsewak Shankar elected president.

1989 Bouterse rejected peace accord reached by President Shankar with guerrilla insurgents, vowed to continue fighting.

1990 Shankar deposed in army coup.

1991 Johan Kraag became interim president. New Front for Democracy won assembly majority. Ronald Venetiaan elected president.

1992 Peace accord with guerrilla groups.

Swaziland
Kingdom of
(*Umbuso weSwatini*)

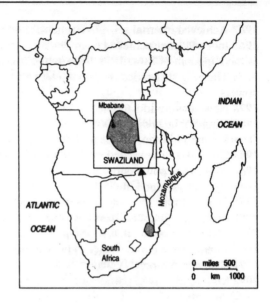

area 17,400 sq km/6,716 sq mi
capital Mbabane
towns Manzini, Big Bend
physical central valley; mountains in W (Highveld); plateau in E
(Lowveld and Lubombo plateau)
head of state and government King Mswati III from 1986
political system near-absolute monarchy
exports sugar, canned fruit, wood pulp, asbestos
currency lilangeni
population (1990 est) 779,000; growth rate 3% p.a.
life expectancy men 47, women 54 (1989)
languages Swazi 90%, English (both official)
religions Christian 57%, animist
literacy men 70%, women 66% (1985 est)
GNP $539 million; $750 per head (1987)
chronology
1903 Swaziland became a special High Commission territory.

1967 Achieved internal self-government.
1968 Independence achieved from Britain, within the Commonwealth, as the Kingdom of Swaziland, with King Sobhuza II as head of state.
1973 The king suspended the constitution and assumed absolute powers.
1978 New constitution adopted.
1982 King Sobhuza died; his place was taken by one of his wives, Dzeliwe, until his son, Prince Makhosetive, reached the age of 21.
1983 Queen Dzeliwe ousted by another wife, Ntombi.
1984 After royal power struggle, it was announced that the crown prince would become king at 18.
1986 Crown prince formally invested as King Mswati III.
1987 Power struggle developed between advisory council Liqoqo and Queen Ntombi over accession of king. Mswati dissolved parliament; new government elected with Sotsha Dlamini as prime minister.
1991 Calls for democratic reform.
1992 Mswati dissolved parliament, assuming 'executive powers'.

Sweden
Kingdom of
(*Konungariket
Sverige*)

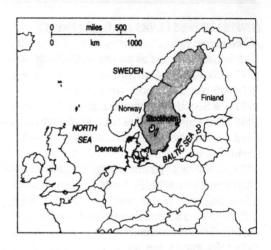

area 450,000 sq km/173,745 sq mi
capital Stockholm
towns Göteborg, Malmö, Uppsala, Norrköping, Västerås
physical mountains in W; plains in S; thickly forested; more than
20,000 islands off the Stockholm coast
environment of the country's 90,000 lakes, 20,000 are affected by acid
rain; 4,000 are so severely acidified that no fish are thought to survive
in them
head of state King Carl XVI Gustaf from 1973
head of government Carl Bildt from 1991
political system constitutional monarchy
exports aircraft, vehicles, missiles, electronics, petrochemicals,
textiles, furnishings, ornamental glass, paper, iron and steel
currency krona
population (1990 est) 8,407,000 (including 17,000 Saami (Lapps) and
1.2 million postwar immigrants from Finland, Turkey, Yugoslavia,
Greece, Iran, other Nordic countries); growth rate 0.1% p.a.
life expectancy men 74, women 81 (1989)
languages Swedish; there are Finnish- and Saami-speaking minorities
religion Lutheran (official) 95%
literacy 99% (1989)

GNP $179 bn; $11,783 per head (1989)
chronology
12th century United as an independent nation.
1397–1520 Under Danish rule.
1914–45 Neutral in both world wars.
1951–76 Social Democratic Labour Party (SAP) in power.
1969 Olof Palme became SAP leader and prime minister.
1971 Constitution amended, creating a single-chamber Riksdag, the governing body.
1975 Monarch's last constitutional powers removed.
1976 Thorbjörn Fälldin, leader of the Centre Party, became prime minister, heading centre-right coalition.
1982 SAP, led by Palme, returned to power.
1985 SAP formed minority government, with Communist support.
1986 Olof Palme murdered. Ingvar Carlsson became prime minister and SAP party leader.
1988 SAP re-elected with reduced majority; Green Party gained representation in Riksdag.
1990 SAP government resigned. Sweden to apply for European Community (EC) membership.
1991 Formal application for EC membership submitted.
Election defeat for SAP; Carlsson resigned. Coalition government formed; Carl Bildt became new prime minister.
1992 Cross-party agreement to solve economic problems.

Switzerland
Swiss Confederation
(German *Schweiz*,
French *Suisse*,
Romansch *Svizzera*)

area 41,300 sq km/15,946 sq mi
capital Bern
towns Zürich, Geneva, Lausanne, Basel
physical most mountainous country in Europe (Alps and Jura
mountains); highest peak Dufourspitze 4,634 m/15,203 ft
environment an estimated 43% of coniferous trees, particularly in the
central alpine region, have been killed by acid rain, 90% of which
comes from other countries. Over 50% of bird species are classified as
threatened
head of state and government Adolf Ogi from 1993
government federal democratic republic
exports electrical goods, chemicals, pharmaceuticals, watches,
precision instruments, confectionery
currency Swiss franc
population (1990 est) 6,628,000; growth rate 0.2% p.a.
life expectancy men 74, women 82 (1989)
languages German 65%, French 18%, Italian 12%, Romansch 1% (all
official)

religions Roman Catholic 50%, Protestant 48%
literacy 99% (1989)
GNP $111 bn (1988); $26,309 per head (1987)
chronology
1648 Became independent of the Holy Roman Empire.
1798–1815 Helvetic Republic established by French revolutionary armies.
1847 Civil war resulted in greater centralization.
1874 Principle of the referendum introduced.
1971 Women given the vote in federal elections.
1984 First female cabinet minister appointed.
1986 Referendum rejected proposal for membership of United Nations.
1989 Referendum supported abolition of citizen army and military service requirements.
1991 18-year-olds allowed to vote for first time in national elections. Four-party coalition remained in power.
1992 René Felber elected president with Adolf Ogi as vice president. Closer ties with European Community (EC) rejected in national referendum.
1993 Ogi replaced Felber as head of state.

Syria
Syrian Arab
Republic
(*al-Jamhuriya al-
Arabya as-Suriya*)

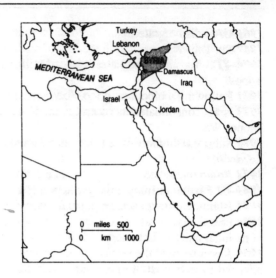

area 185,200 sq km/71,506 sq mi
capital Damascus
towns Aleppo, Homs, Hama, Latakia
physical mountains alternate with fertile plains and desert areas;
Euphrates River
head of state and government Hafez al-Assad from 1971
political system socialist republic
exports cotton, cereals, oil, phosphates, tobacco
currency Syrian pound
population (1990 est) 12,471,000; growth rate 3.5% p.a.
life expectancy men 67, women 69 (1989)
languages Arabic 89% (official), Kurdish 6%, Armenian 3%
religions Sunni Muslim 74%; ruling minority Alawite, and other
Islamic sects 16%; Christian 10%
literacy men 76%, women 43% (1985 est)
GNP $17 bn (1986); $702 per head
chronology
1946 Achieved full independence from France.
1958 Merged with Egypt to form the United Arab Republic (UAR).

1961 UAR disintegrated.

1967 Six-Day War resulted in the loss of territory to Israel.

1970–71 Syria supported Palestinian guerrillas against Jordanian troops.

1971 Following a bloodless coup, Hafez al-Assad became president.

1973 Israel consolidated its control of the Golan Heights after the Yom Kippur War.

1976 Substantial numbers of troops committed to the civil war in Lebanon.

1978 Assad re-elected.

1981–82 Further military engagements in Lebanon.

1982 Islamic militant uprising suppressed; 5,000 dead.

1984 Presidents Assad and Gemayel approved plans for government of national unity in Lebanon.

1985 Assad secured the release of 39 US hostages held in an aircraft hijacked by extremist Shi'ite group, Hezbollah. Assad re-elected.

1987 Improved relations with USA and attempts to secure the release of Western hostages in Lebanon.

1989 Diplomatic relations with Morocco restored. Continued fighting in Lebanon; Syrian forces reinforced in Lebanon; diplomatic relations with Egypt restored.

1990 Diplomatic relations with Britain restored.

1991 Syria fought against Iraq in Gulf War. President Assad agreed to US Middle East peace plan. Assad re-elected as president.

Taiwan
Republic of China
(*Chung Hua Min Kuo*)

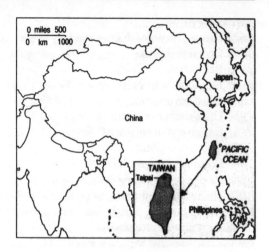

area 36,179 sq km/13,965 sq mi
capital Taipei
towns Kaohsiung, Keelung
physical island (formerly Formosa) off People's Republic of China; mountainous, with lowlands in W
environment industrialization has taken its toll: an estimated 30% of the annual rice crop is dangerously contaminated with mercury, cadmium, and other heavy metals
head of state Lee Teng-hui from 1988
head of government Lien Chan from 1993
political system emergent democracy
exports textiles, steel, plastics, electronics, foodstuffs
currency New Taiwan dollar
population (1990) 20,454,000 (Taiwanese 84%, mainlanders 14%); growth rate 1.4% p.a.
life expectancy 70 men, 75 women (1986)
languages Mandarin Chinese (official); Taiwan, Hakka dialects
religions officially atheist; Taoist, Confucian, Buddhist, Christian
literacy 90% (1988)
GNP $119.1 bn; $6,200 per head (1988)

chronology

1683 Taiwan (Formosa) annexed by China.

1895 Ceded to Japan.

1945 Recovered by China.

1949 Flight of Nationalist government to Taiwan after Chinese communist revolution.

1954 US-Taiwanese mutual defence treaty.

1971 Expulsion from United Nations.

1972 Commencement of legislature elections.

1975 President Chiang Kai-shek died; replaced as Kuomintang leader by his son, Chiang Ching-kuo.

1979 USA severed diplomatic relations and annulled 1954 security treaty.

1986 Democratic Progressive Party (DPP) formed as opposition to the nationalist Kuomintang.

1987 Martial law lifted; opposition parties legalized; press restrictions lifted.

1988 President Chiang Ching-kuo died; replaced by Taiwanese-born Lee Teng-hui.

1989 Kuomintang won assembly elections.

1990 Formal move towards normalization of relations with China. Hau Pei-tsun became prime minister.

1991 President Lee Teng-hui declared end to state of civil war with China. Constitution amended. Kuomintang won landslide victory in assembly elections.

1992 Diplomatic relations with South Korea broken. Dec: in first fully democratic elections Kuomintang lost support to DPP but still secured a majority of seats.

1993 Lien Chan appointed prime minister.

Tajikistan
Republic of

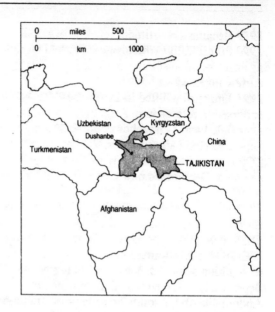

area 143,100 sq km/55,251 sq mi
capital Dushanbe
towns Khodzhent (formerly Leninabad), Kurgan-Tyube, Kulyab
physical mountainous, more than half of its territory lying above 3,000 m/10,000 ft; huge mountain glaciers, which are the source of many rapid rivers
head of state Akbasho Iskandrov from 1992
head of government Abdumalik Abdulojonov from 1992
political system emergent democracy
products fruit, cereals, cotton, cattle , sheep, silks, carpets, coal, lead, zinc, chemicals, oil, gas
population (1990) 5,300,000 (Tajik 63%, Uzbek 24%, Russian 8%, Tatar 1%, Kyrgyz 1%, Ukrainian 1%)
language Tajik, similar to Farsi (Persian)
religion Sunni Muslim
chronology
1921 Part of Turkestan Soviet Socialist Autonomous Republic.

1929 Became a constituent republic of USSR.

1990 Ethnic Tajik/Armenian conflict in Dushanbe resulted in rioting against Communist Party of Tajikistan (TCP); state of emergency and curfew imposed.

1991 Jan: curfew lifted in Dushanbe. March: maintenance of Union endorsed in referendum. Aug: President Makhkamov forced to resign after failed anti-Gorbachev coup; TCP broke links with Moscow. Sept: declared independence; Rakhman Nabiyev elected president; TCP renamed Socialist Party of Tajikistan (SPT); state of emergency declared. Dec: joined new Commonwealth of Independent States (CIS).

1992 Jan: admitted into Conference for Security and Cooperation in Europe (CSCE). Nabiyev temporarily ousted; state of emergency lifted. Feb: joined the Muslim Economic Cooperation Organization (ECO). March: admitted into United Nations (UN); US diplomatic recognition achieved. May: coalition government formed. Sept: Nabiyev forced to resign; replaced by Akbasho Iskandrov; Abdumalik Abdulojonov became prime minister.

1993 Civil war between the forces of the country's communist former rulers and Islamic and pro-democracy groups continued to rage.

Tanzania
United Republic of
(*Jamhuri ya
Muungano wa
Tanzania*)

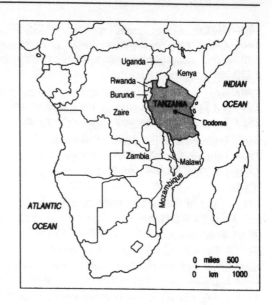

area 945,000 sq km/364,865 sq mi
capital Dodoma (since 1983)
towns Zanzibar Town, Mwanza, Dar es Salaam
physical central plateau; lakes in N and W; coastal plains; lakes
Victoria, Tanganyika, and Niasa
environment the black rhino faces extinction as a result of poaching
head of state and government Ali Hassan Mwinyi from 1985
political system one-party socialist republic
exports coffee, cotton, sisal, cloves, tea, tobacco, cashew nuts,
diamonds
currency Tanzanian shilling
population (1990 est) 26,070,000; growth rate 3.5% p.a.
life expectancy men 49, women 54 (1989)
languages Kiswahili, English (both official)
religions Muslim 35%, Christian 35%, traditional 30%
literacy 85% (1987)
GNP $4.9 bn; $258 per head (1987)

chronology
16th–17th centuries Zanzibar under Portuguese control.
1890–1963 Zanzibar became a British protectorate.
1920–46 Tanganyika administered as a British League of Nations mandate.
1946–62 Tanganyika came under United Nations (UN) trusteeship.
1961 Tanganyika achieved independence from Britain, within the Commonwealth, with Julius Nyerere as prime minister.
1962 Tanganyika became a republic with Nyerere as president.
1964 Tanganyika and Zanzibar became the United Republic of Tanzania with Nyerere as president.
1967 East African Community (EAC) formed. Nyerere committed himself to building a socialist state through a series of village cooperatives (the Arusha Declaration).
1977 Revolutionary Party of Tanzania (CCM) proclaimed the only legal party. EAC dissolved.
1978 Ugandan forces repulsed after crossing into Tanzania.
1979 Tanzanian troops sent to Uganda to help overthrow the president, Idi Amin.
1985 Nyerere retired from presidency but stayed on as CCM leader; Ali Hassan Mwinyi became president.
1990 Nyerere surrendered CCM leadership; replaced by President Mwinyi.
1992 CCM agreed to abolish one-party rule. East African cooperation pact with Kenya and Uganda to be re-established.

Thailand
Kingdom of
(*Prathet Thai* or
Muang-Thai)
(formerly *Siam* to
1939 and 1945–49)

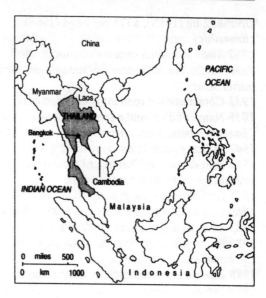

area 513,115 sq km/198,108 sq mi
capital Bangkok
towns Chiangmai, Nakhon Sawan
physical mountainous, semi-arid plateau in NE, fertile central region,
tropical isthmus in S
environment tropical rainforest was reduced to 18% of the land area
1988 (from 93% in 1961); logging banned by the government 1988
head of state King Bhumibol Adulyadej from 1946
head of government Chuan Leekpai from 1992
political system military-controlled emergent democracy
exports rice, textiles, rubber, tin, rubies, sapphires, maize, tapioca
currency baht
population (1990 est) 54,890,000 (Thai 75%, Chinese 14%); growth
rate 2% p.a.
life expectancy men 62, women 68 (1989)
languages Thai and Chinese (both official); regional dialects
religions Buddhist 95%, Muslim 4%
literacy 89% (1988)

GNP $52 bn (1988); $771 per head (1988)
chronology
1782 Siam absolutist dynasty commenced.
1896 Anglo-French agreement recognized Siam as an independent buffer state.
1932 Constitutional monarchy established.
1939 Name of Thailand adopted.
1941–44 Japanese occupation.
1947 Military seized power in coup.
1972 Withdrawal of Thai troops from South Vietnam.
1973 Military government overthrown.
1976 Military reassumed control.
1980 General Prem Tinsulanonda assumed power.
1983 Civilian government formed; martial law maintained.
1988 Prime Minister Prem resigned; replaced by Chatichai Choonhavan.
1989 Thai pirates continued to murder, pillage, and kidnap Vietnamese 'boat people' at sea.
1991 Military seized power in coup. Interim civilian government formed under Anand Panyarachun. 50,000 demonstrated against new military-oriented constitution.
1992 March: general election produced five-party coalition; Narong Wongwan named premier but removed a month later. April: appointment of General Suchinda Kraprayoon as premier provoked widespread riots. May: Suchinda forced to stand down. June: Anand made interim prime minister. Sept: new coalition government led by Chuan Leekpai.

Togo
Republic of
(*République
Togolaise*)

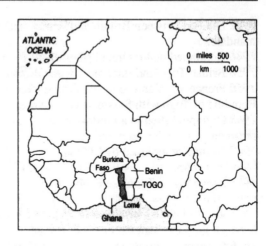

area 56,800 sq km/21,930 sq mi
capital Lomé
towns Sokodé, Kpalimé
physical two savanna plains, divided by range of hills NE–SW; coastal lagoons and marsh
environment the homes of thousands of people in Keto were destroyed by coastal erosion as a result of the building of the Volta dam
head of state Etienne Gnassingbé Eyadéma from 1967
head of government Jospeh Kokou Koffigoh from 1991
political system transitional
exports phosphates, cocoa, coffee, coconuts
currency franc CFA
population (1990 est) 3,566,000; growth rate 3% p.a.
life expectancy men 53, women 57 (1989)
languages French (official), Ewe, Kabre
religions animist 46%, Catholic 28%, Muslim 17%, Protestant 9%
literacy men 53%, women 28% (1985 est)
GNP $1.3 bn (1987); per head (1985)
chronology
1885–1914 Togoland was a German protectorate until captured by Anglo-French forces.

1922 Divided between Britain and France under League of Nations mandate.

1946 Continued under United Nations trusteeship.

1956 British Togoland integrated with Ghana.

1960 French Togoland achieved independence from France as the Republic of Togo with Sylvanus Olympio as head of state.

1963 Olympio killed in a military coup. Nicolas Grunitzky became president.

1967 Grunitzky replaced by Lt-Gen Etienne Gnassingbé Eyadéma in bloodless coup.

1973 Assembly of Togolese People (RPT) formed as sole legal political party.

1975 Lomé convention signed in Lomé, establishing trade links between the European Community (EC) and developing countries.

1979 Eyadéma returned in election. Further EC Lomé convention signed.

1986 Attempted coup failed.

1991 Eyadéma legalized opposition parties. National conference elected Joseph Kokou Koffigoh head of interim government; troops loyal to Eyadéma failed to reinstate him.

1992 Overwhelming referendum support for multiparty politics.

1993 Feb: all-party talks to avoid civil war began in France but were suspended after disagreements among participants.

Tonga
 Kingdom of
(*Pule'anga Fakatu'i
'o Tonga*)
or *Friendly Islands*

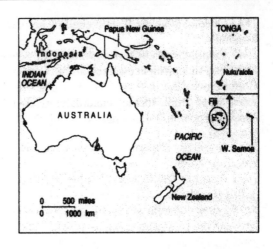

area 750 sq km/290 sq mi
capital Nuku'alofa (on Tongatapu island)
towns Pangai, Neiafu
physical three groups of islands in SW Pacific, mostly coral
formations, but actively volcanic in W
head of state King Taufa'ahau Tupou IV from 1965
head of government Baron Vaea from 1991
political system constitutional monarchy
currency Tongan dollar or pa'anga
population (1988) 95,000; growth rate 2.4% p.a.
life expectancy men 69, women 74 (1989)
languages Tongan (official), English
religions Wesleyan 47%, Roman Catholic 14%, Free Church of Tonga
14%, Mormon 9%, Church of Tonga 9%
literacy 93% (1988)
GNP $65 million (1987); $430 per head
chronology
1831 Tongan dynasty founded by Prince Taufa'ahau Tupou.
1900 Became a British protectorate.
1958 Treaty with Britain brought Tonga increased control over its
internal affairs.

1965 Queen Salote died; succeeded by her son, King Taufa'ahau Tupou IV.
1967 Revised treaty with Britain brought greater independence.
1970 Full independence achieved from Britain within the Commonwealth.
1990 Three prodemocracy candidates elected. Calls for reform of absolutist power.

Trinidad and Tobago
Republic of

area Trinidad 4,828 sq km/1,864 sq mi and Tobago 300 sq km/
116 sq mi
capital Port-of-Spain
towns San Fernando, Arima, Scarborough (Tobago)
physical comprises two main islands and some smaller ones; coastal
swamps and hills E–W
head of state Noor Hassanali from 1987
head of government Patrick Manning from 1991
political system democratic republic
exports oil, petroleum products, chemicals, sugar, cocoa
currency Trinidad and Tobago dollar
population (1990 est) 1,270,000 (African descent 40%, Indian 40%,
European 16%, Chinese 2% and others), 1.2 million on Trinidad;
growth rate 1.6% p.a.
life expectancy men 68, women 72 (1989)
languages English (official), Hindi, French, Spanish
religions Roman Catholic 32%, Protestant 29%, Hindu 25%,
Muslim 6%
literacy 97% (1988)

GNP $4.5 bn; $3,731 per head (1987)
chronology
1888 Trinidad and Tobago united as a British colony.
1956 People's National Movement (PNM) founded.
1959 Achieved internal self-government, with PNM leader Eric Williams as chief minister.
1962 Independence achieved from Britain, within the Commonwealth, with Williams as prime minister.
1976 Became a republic, with Ellis Clarke as president and Williams as prime minister.
1981 Williams died and was succeeded by George Chambers, with Arthur Robinson as opposition leader.
1986 National Alliance for Reconstruction (NAR), headed by Arthur Robinson, won general election.
1987 Noor Hassanali became president.
1990 Attempted antigovernment coup defeated.
1991 General election saw victory for PNM, with Patrick Manning as prime minister.

Tunisia
Tunisian Republic
(*al-Jumhuriya
at-Tunisiya*)

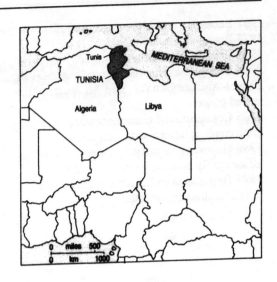

area 164,150 sq km/63,378 sq mi
capital Tunis
towns ports Sfax, Sousse, Bizerta
physical arable and forested land in N graduates towards desert in S
head of state and government Zine el-Abidine Ben Ali from 1987
political system emergent democratic republic
exports oil, phosphates, chemicals, textiles, food, olive oil
currency dinar
population (1990 est) 8,094,000; growth rate 2% p.a.
life expectancy men 68, women 71 (1989)
languages Arabic (official), French
religion Sunni Muslim 95%; Jewish, Christian
literacy men 68%, women 41% (1985 est)
GNP $9.6 bn (1987); $1,163 per head (1986)
chronology
1883 Became a French protectorate.
1955 Granted internal self-government.
1956 Independence achieved from France as a monarchy, with Habib
Bourguiba as prime minister.

1957 Became a republic with Bourguiba as president.

1975 Bourguiba made president for life.

1985 Diplomatic relations with Libya severed.

1987 Bourguiba removed Prime Minister Rashed Sfar and appointed Zine el-Abidine Ben Ali. Ben Ali declared Bourguiba incompetent and seized power.

1988 Constitutional changes towards democracy announced. Diplomatic relations with Libya restored.

1989 Government party, Constitutional Democratic Rally (RDC), won all assembly seats in general election.

1991 Opposition to US actions during the Gulf War. Crackdown on religious fundamentalists.

Turkey
Republic of
(*Türkiye
Cumhuriyeti*)

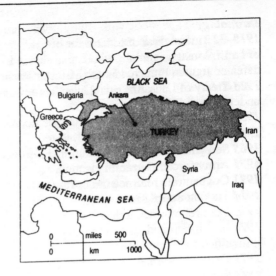

area 779,500 sq km/300,965 sq mi
capital Ankara
towns Istanbul, Izmir
physical central plateau surrounded by mountains
environment only 0.3% of the country is protected by national parks
and reserves compared with a global average of 7% per country
head of state to be appointed
head of government Suleyman Demirel from 1991
political system democratic republic
exports cotton, yarn, hazelnuts, citrus fruits, tobacco, dried fruit,
chromium ores
currency Turkish lira
population (1990 est) 56,549,000 (Turkish 85%, Kurdish 12%);
growth rate 2.1% p.a.
life expectancy men 63, women 66 (1989)
languages Turkish (official), Kurdish, Arabic
religion Sunni Muslim 98%
literacy men 86%, women 62% (1985)
GNP $62 bn (1987); $1,160 per head (1986)

chronology

1919–22 Turkish War of Independence provoked by Greek occupation of Izmir. Mustafa Kemal (Atatürk), leader of nationalist congress, defeated Italian, French, and Greek forces.

1923 Treaty of Lausanne established Turkey as independent republic under Kemal. Westernization began.

1950 First free elections; Adnan Menderes became prime minister.

1960 Menderes executed after military coup by General Cemal Gürsel.

1965 Suleyman Demirel became prime minister.

1971 Army forced Demirel to resign.

1973 Civilian rule returned under Bulent Ecevit.

1974 Turkish troops sent to protect Turkish community in Cyprus.

1975 Demirel returned to head of a right-wing coalition.

1978 Ecevit returned, as head of coalition, in the face of economic difficulties and factional violence.

1979 Demeril returned. Violence grew.

1980 Army took over, and Bulent Ulusu became prime minister. Harsh repression of political activists attracted international criticism.

1982 New constitution adopted.

1983 Ban on political activity lifted. Turgut Özal became prime minister.

1987 Özal maintained majority in general election.

1988 Improved relations and talks with Greece.

1989 Turgut Özal elected president; Yildirim Akbulut became prime minister. Application to join European Community (EC) rejected.

1991 Mesut Yilmaz became prime minister. Turkey sided with UN coalition against Iraq in Gulf War. Conflict with Kurdish minority continued. Coalition government formed under Suleyman Demirel after inconclusive election result.

1992 Earthquake claimed thousands of lives.

1993 April: Turgut Özal died of heart attack.

Turkmenistan
Republic of

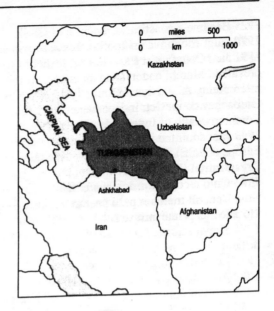

area 488,100 sq km/188,406 sq mi
capital Ashkhabad
towns Chardzhov, Mary (Merv), Nebit-Dag, Krasnovodsk
physical some 90% of land is desert including the Kara Kum 'Black Sands' desert (area 310,800 sq km/120,000 sq mi)
head of state Saparmurad Niyazov from 1991
head of government Sakhat Muradov from 1992
political system socialist pluralist
products silk, karakul, sheep, astrakhan fur, carpets, chemicals, rich deposits of petroleum, natural gas, sulphur, and other industrial raw materials
population (1990) 3,600,000 (Turkmen 72%, Russian 10%, Uzbek 9%, Kazakh 3%, Ukrainian 1%)
language West Turkic, closely related to Turkish
religion Sunni Muslim
chronology
1921 Part of Turkestan Soviet Socialist Autonomous Republic.

1925 Became a constituent republic of USSR.

1990 Aug: economic and political sovereignty declared.

1991 Jan: Communist Party leader Sakhat Niyazov became state president. March: endorsed maintenance of the Union in USSR referendum. Aug: Niyazov initially supported attempted anti-Gorbachev coup. Oct: independence declared. Dec: joined new Commonwealth of Independent States (CIS).

1992 Jan: admitted into Conference for Security and Cooperation in Europe (CSCE). Feb: joined the Muslim Economic Cooperation Organization (ECO). March: admitted into United Nations (UN); US diplomatic recognition achieved. May: new constitution adopted. Nov–Dec: 60-member parliament popularly elected with Sakhat Muradov as prime minister.

Tuvalu
South West Pacific
State of
(formerly *Ellice
Islands*)

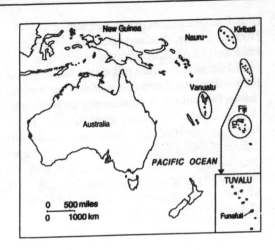

area 25 sq km/9.5 sq mi
capital Funafuti
physical nine low coral atolls forming a chain of 579 km/650 mi in the
SW Pacific
head of state Elizabeth II from 1978, represented by governor general
head of government Bikenibeu Paeniu from 1989
political system liberal democracy
exports copra, handicrafts, stamps
currency Australian dollar
population (1990 est) 9,000 (Polynesian 96%); growth rate 3.4% p.a.
life expectancy 60 men, 63 women (1989)
languages Tuvaluan, English
religion Christian (Protestant)
literacy 96% (1985)
GDP (1983) $711 per head
chronology
1892 Became a British protectorate forming part of the Gilbert and
Ellice Islands group.
1916 The islands acquired colonial status.
1975 The Ellice Islands were separated from the Gilbert Islands.
1978 Independence achieved from Britain, within the Commonwealth,

with Toaripi Lauti as prime minister.

1981 Dr Tomasi Puapua replaced Lauti as premier; country reverted to former name Tuvalu.

1986 Islanders rejected proposal for republican status.

1989 Bikenibeu Paeniu elected new prime minister and pledged to reduce the country's dependence on foreign aid.

Uganda
Republic of

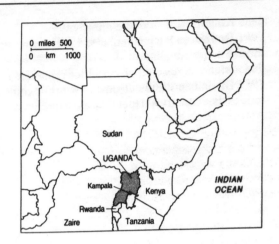

area 236,600 sq km/91,351 sq mi
capital Kampala
towns Jinja, M'Bale, Entebbe, Masaka
physical plateau with mountains in W; forest and grassland; arid in NE
head of state and government Yoweri Museveni from 1986
political system emergent democratic republic
exports coffee, cotton, tea, copper
currency Uganda new shilling
population (1990 est) 17,593,000 (largely the Baganda, after whom the country is named; also Langi and Acholi, some surviving Pygmies); growth rate 3.3% p.a.
life expectancy men 49, women 51 (1989)
languages English (official), Kiswahili, Luganda, and other African languages
religions Roman Catholic 33%, Protestant 33%, Muslim 16%, animist
literacy men 70%, women 33% (1985 est)
GNP $3.6 bn (1987); $220 (1985 est)
chronology
1962 Independence achieved from ... with Milton Obote as prime minister
1963 Proclaimed a federal republic within the Commonwealth,
... II as president.

1966 King Mutesa ousted in coup led by Obote, who ended the federal status and became executive president.

1969 All opposition parties banned after an assassination attempt on Obote.

1971 Obote overthrown in army coup led by Maj-Gen Idi Amin Dada; ruthlessly dictatorial regime established; nearly 49,000 Ugandan Asians expelled; over 300,000 opponents of regime killed.

1978 Amin forced to leave country by opponents backed by Tanzanian troops. Provisional government set up with Yusuf Lule as president. Lule replaced by Godfrey Binaisa.

1978–79 Fighting broke out against Tanzanian troops.

1980 Binaisa overthrown by army. Elections held and Milton Obote returned to power.

1985 After opposition by National Resistance Army (NRA), and indiscipline in army, Obote ousted by Brig Tito Okello; power-sharing agreement entered into with NRA leader Yoweri Museveni.

1986 Agreement ended; Museveni became president, heading broad-based coalition government.

1992 Announcement made that East African cooperation pact with Kenya and Tanzania would be revived.

Ukraine

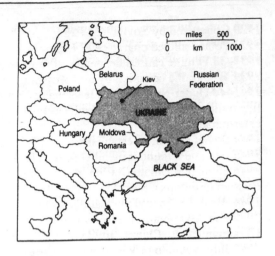

area 603,700 sq km/233,089 sq mi
capital Kiev
towns Kharkov, Donetsk, Odessa, Dnepropetrovsk, Lviv (Lvov),
Krivoi Rog, Zaporozhye
physical Russian plain; Carpathian and Crimean Mountains; rivers:
Dnieper (with the Dnieper dam 1932), Donetz, Bug
head of state Leonid Kravchuk from 1990
head of government Leonid Kuchma from 1992
political system emergent democracy
produce grain, coal, oil, various minerals
currency δ ma
population (1   Ru51,800,000 (Ukrainian 73%, Russian 22%,
Byelorussian,   speaking Jews 1%—some 1.5 million have
emigrated to the US   (S 000 to Canada)
language Ukrainian   with a literature that goes back to the
Middle Ages
religions traditionally Ukrain   ox; also Ukrainian Catholic
chronology
1918 Independent People's Republ   d.

1920 Conquered by Soviet Red Army.

1921 Poland alloted charge of W Ukraine.

1932–33 Famine caused the deaths of more than 7.5 million people.

1939 W Ukraine occupied by Red Army.

1941–44 Under Nazi control; Jews massacred at Babi Yar; more than five million Ukrainians and Ukrainian Jews deported and exterminated.

1944 Soviet control re-established.

1945 Became a founder member of the United Nations.

1946 Ukrainian Uniate Church proscribed and forcibly merged with Russian Orthodox Church.

1986 April: Chernobyl nuclear disaster.

1989 Feb: Ukrainian People's Movement (Rukh) established. Ban on Ukrainian Uniate Church lifted.

1990 July: voted to proclaim sovereignty; Leonid Kravchuk indirectly elected as president; sovereignty declared.

1991 Aug: demonstrations against the abortive anti-Gorbachev coup; independence declared, pending referendum; Communist Party (CP) activities were suspended. Oct: voted to create independent army. Dec: Kravchuk popularly elected president; independence overwhelmingly endorsed in referendum; joined new Commonwealth of Independent States (CIS); independence acknowledged by USA and European Community (EC).

1992 Jan: admitted into Conference on Security and Cooperation in Europe (CSCE); pipeline deal with Iran to end dependence on Russian oil; prices freed. Feb: prices 'temporarily' re-regulated. March: agreed tactical arms-shipments to Russia suspended. May: Crimean sovereignty declared, but subsequently rescinded. Aug: joint control of Black Sea fleet agreed with Russia. Oct: Leonid Kuchma became prime minister. Production declined by 20% during ... budget deficit at

1993 Inflation at 35% a month in early part ... 44% of GDP.

United Arab Emirates
(UAE)
(*Ittihad al-Imarat al-Arabiyah*)
federation of the emirates of Abu Dhabi, Ajman, Dubai, Fujairah, Ras al Khaimah, Sharjah, Umm al Qaiwain

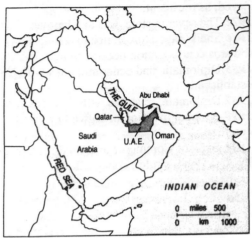

total area 83,657 sq km/32,292 sq mi
capital Abu Dhabi
towns Dubai
physical desert and flat coastal plain; mountains in E
head of state and of government Sheik Sultan Zayed bin al-Nahayan of Abu Dhabi from 1971
political system absolutism
exports oil, natural gas, fish, dates
currency UAE dirham
population (1990 est) 2,250,000 (10% nomadic); growth rate 6.1% p.a.
life expectancy men 68, women 72 (1989)
languages Arabic (official), Farsi, Hindi, Urdu, English
religions 68% 96%, Christian, Hindu
literacy 68% 96%, Christian, Hindu
GNP $22 bn; $11,900 per head
chronology
1952 Trucial C
1971 Federation
States formed Un shed.
tes formed; later dissolved. Six Trucial
tes, with ruler of Abu Dhabi, Sheik

Zayed, as president.
1972 The seventh state, Ras al Khaimah, joined the federation.
1976 Sheik Zayed threatened to relinquish presidency unless progress towards centralization became more rapid.
1985 Diplomatic and economic links with USSR and China established.
1987 Diplomatic relations with Egypt restored.
1990–91 Iraqi invasion of Kuwait opposed; UAE fights with the UN coalition.
1991 Bank of Commerce and Credit International (BCCI), controlled by Abu Dhabi's ruler, collapsed.

United Kingdom
of Great Britain and
Northern Ireland
(UK)

area 244,100 sq km/94,247 sq mi
capital London
towns Birmingham, Glasgow, Leeds, Liverpool, Manchester, Belfast,
Cardiff
physical became separated from European continent about 6000 BC;
rolling landscape, increasingly mountainous towards the N, with
Grampian Mountains in Scotland, Pennines in N England, Cambrian
Mountains in Wales; rivers include Thames, Severn, and Spey
territories Anguilla, Bermuda, British Antarctic Territory, British
Indian Ocean Territory, British Virgin Islands, Cayman Islands,
Falkland Islands, Gibraltar, Hong Kong (until 1997), Montserrat,
Pitcairn Islands, Turks and Caicos and Dependencies (Ascension, Tristan da
Cunha), an estimated lands
environment forests have been damaged
head of state Elizabeth II from the highest percentage in Europe) of
head of government John Major in
political system liberal democracy

exports cereals, rape, sugar beet, potatoes, meat and meat products, poultry, dairy products, electronic and telecommunications equipment, engineering equipment and scientific instruments, oil and gas, petrochemicals, pharmaceuticals, fertilizers, film and television programmes, aircraft

currency pound sterling

population (1990 est) 57,121,000 (English 81.5%, Scottish 9.6%, Welsh 1.9%, Irish 2.4%, Ulster 1.8%); growth rate 0.1% p.a.

religion Christian (Protestant 55%, Roman Catholic 10%); Muslim, Jewish, Hindu, Sikh

life expectancy men 72, women 78 (1989)

languages English, Welsh, Gaelic

literacy 99% (1989)

GNP $758 bn; $13,329 per head (1988) 7z

recent chronology

1945 Labour government under Clement Attlee; welfare state established.

1951 Conservatives under Winston Churchill defeated Labour.

1956 Suez Crisis.

1964 Labour victory under Harold Wilson.

1970 Conservatives under Edward Heath defeated Labour.

1972 Parliament prorogued in Northern Ireland; direct rule from Westminster began.

1973 UK joined European Economic Community.

1974 Three-day week, coal strike; Wilson replaced Heath.

1976 James Callaghan replaced Wilson as prime minister.

1977 Liberal–Labour pact.

1979 Victory for Conservatives under Margaret Thatcher.

1981 Formation of Social Democratic Party (SDP). Riots occurred in inner cities.

1982 Unemployment over 3 million. Falklands War.

1983 Thatcher re-elected.

1984–85 Coal strike, the longest in British history.

1986 Abolition of metropolitan counties.

1987 Thatcher re-elected for third term.

1988 Liberals and most of SDP merged into the Social and Liberal

Democrats, leaving a splinter SDP. Inflation and interest rates rose.
1989 The Green Party polled 2 million votes in the European elections.
1990 Riots as poll tax introduced in England. Troops sent to the
Persian Gulf following Iraq's invasion of Kuwait. British hostages
held in Iraq, later released. Britain joined European exchange rate
mechanism (ERM). Thatcher replaced by John Major as Conservative
leader and prime minister.
1991 British troops took part in US-led war against Iraq under United
Nations umbrella. Support was given to the USSR during the
dissolution of communism and the restoration of independence to the
republics. John Major visited Beijing to sign agreement with China on
new Hong Kong airport. At home, Britain suffered severe economic
recession and rising unemployment.
1992 Economic recession continued. April: Conservative Party, led by
John Major, won fourth consecutive general election, but with reduced
majority. Neil Kinnock resigned as Labour leader. July: John Smith
became new Labour leader. Sept: sterling devalued and UK withdrawn
from ERM. Oct: drastic pit-closure programme encountered massive
public opposition; subsequently reviewed. John Major's popularity at
unprecedentedly low rating. Nov: government motion in favour of
ratification of Maastricht Treaty narrowly passed. Revelations of past
arms sales to Iraq implicated senior government figures, including the
prime minister.

United States of America

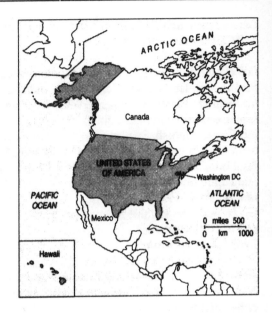

area 9,368,900 sq km/3,618,770 sq mi

capital Washington DC

towns New York, Los Angeles, Chicago, Philadelphia, Detroit, San Francisco

physical topography and vegetation from tropical (Hawaii) to arctic (Alaska); mountain ranges parallel with E and W coasts; the Rocky Mountains separate rivers emptying into the Pacific from those flowing into the Gulf of Mexico; Great Lakes in N; rivers include Hudson, Mississippi, Missouri, Colorado, Columbia, Snake, Rio Grande, Ohio

environment the USA produces the world's largest quantity of municipal waste per person (850 kg/1,900 lb)

territories the commonwealths of Puerto Rico and Northern Marianas; the federated states of Micronesia; Guam, the US Virgin Islands, American Samoa, Wake Island, Midway Islands, Marshall Islands, Belau, and Johnston and Sand Islands

head of state and government Bill Clinton from 1993
political system liberal democracy
currency US dollar
population (1990 est) 250,372,000 (white 80%, black 12%, Asian/
Pacific islander 3%, American Indian, Inuit, and Aleut 1%, Hispanic
(included in above percentages) 9%); growth rate 0.9% p.a.
life expectancy men 72, women 79 (1989)
languages English, Spanish
religions Christian 86.5% (Roman Catholic 26%, Baptist 19%,
Methodist 8%, Lutheran 5%), Jewish 1.8%, Muslim 0.5%, Buddhist
and Hindu less than 0.5%
literacy 99% (1989)
GNP $3,855 bn (1983); $13,451 per head
recent chronology
1945 USA ended war in the Pacific by dropping atom bombs on
Hiroshima and Nagasaki, Japan.
1950–53 US involvement in Korean War. McCarthy anticommunist
investigations (HUAC) became a 'witch hunt'.
1954 Civil Rights legislation began with the ending of the segregation
of black and white students in public schools.
1957 Civil Rights bill on voting.
1958 First US satellite in orbit.
1961 Abortive CIA-backed invasion of Cuba at the Bay of Pigs.
1963 President Kennedy assassinated; L B Johduring nson assumed
the presidency.
1964–68 'Great Society' civil-rights and welfare measures in the
Omnibus Civil Rights bill.
1964–75 US involvement in Vietnam War.
1965 US intervention in Dominican Republic.
1969 US astronaut Neil Armstrong was the first human on the Moon.
1973 OPEC oil embargo almost crippled US industry and consumers.
Inflation began.
1973–74 Watergate scandal began during Richard Nixon's re-election
campaign and ended just before impeachment; Nixon resigned as
president; replaced by Gerald Ford, who 'pardoned' Nixon.
1975 Final US withdrawal from Vietnam.

1979 US–Chinese diplomatic relations normalized.

1981 Ronald Reagan inaugurated as president. Space shuttle mission was successful.

1983 US invasion of Grenada.

1986 'Irangate' scandal over secret US government arms sales to Iran, with proceeds to antigovernment Contra guerrillas in Nicaragua.

1987 Reagan and Gorbachev (for USSR) signed intermediate-range nuclear forces treaty. Wall Street stock-market crash caused by programme trading.

1988 USA became world's largest debtor nation, owing $532 billion. George Bush elected president.

1989 Bush met Gorbachev at Malta, end to Cold War declared; high-level delegation sent to China amid severe criticism; large troop reductions and budget cuts announced for US military; USA invaded Panama; Noriega taken into custody.

1990 Bush and Gorbachev met again. Nelson Mandela freed in South Africa, toured USA. US troops sent to Middle East following Iraq's invasion of Kuwait.

1991 Jan–Feb: US-led assault drove Iraq from Kuwait in Gulf War. US support was given to the USSR during the dissolution of communism and the recognition of independence of the Baltic republics. July: Strategic Arms Reduction Treaty (START) signed at US–Soviet summit in Moscow. Nov: Bush co-hosted Middle East peace conference in Spain.

1992 Bush's popularity slumped as economic recession continued. Widespread riots in Los Angeles. Nov: Bill Clinton won presidential elections for the Democrats; independent candidate Ross Perot won nearly 20% of votes.

1993 Jan: Clinton delayed executive order to suspend ban on homosexuality in the armed forces. Feb: medium-term economic plan proposed by Clinton to Congress to cut federal budget deficit.

Uruguay
Oriental Republic of
(*República Oriental
del Uruguay*)

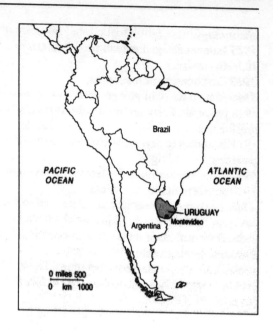

area 176,200 sq km/68,031 sq mi
capital Montevideo
towns Salto, Paysandú
physical grassy plains (pampas) and low hills
head of state and government Luis Lacalle Herrera from 1989
political system democratic republic
exports meat and meat products, leather, wool, textiles
currency nuevo peso
population (1990 est) 3,002,000 (Spanish, Italian; mestizo, mulatto,
black); growth rate 0.7% p.a.
life expectancy men 68, women 75 (1989)
language Spanish
religion Roman Catholic 66%
literacy 96% (1984)
GNP $7.5 bn; $2,470 per head (1988)

chronology

1825 Independence declared from Brazil.

1836 Civil war.

1930 First constitution adopted.

1966 Blanco party in power, with Jorge Pacheco Areco as president.

1972 Colorado Party returned, with Juan Maria Bordaberry Arocena as president.

1976 Bordaberry deposed by army; Dr Méndez Manfredini became president.

1984 Violent antigovernment protests after ten years of repressive rule.

1985 Agreement reached between the army and political leaders for return to constitutional government. Colorado Party won general election; Dr Julio Maria Sanguinetti became president.

1986 Government of national accord established under President Sanguinetti's leadership.

1989 Luis Lacalle Herrera elected president.

Uzbekistan
Republic of

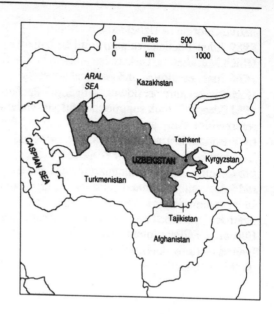

area 447,400 sq km/172,741 sq mi
capital Tashkent
towns Samarkand, Bukhara, Namangan
physical oases in the deserts; rivers: Amu Darya, Syr Darya; Fergana
Valley; rich in mineral deposits
head of state Islam Karimov from 1990
head of government Abdul Hashim Mutalov from 1991
political system socialist pluralist
products rice, dried fruit, vines (all grown by irrigation); cotton, silk
population (1990) 20,300,000 (Uzbek 71%, Russian 8%, Tajik 5%,
Kazakh 4%)
language Uzbek, a Turkic language
religion Sunni Muslim
chronology
1921 Part of Turkestan Soviet Socialist Autonomous Republic.
1925 Became constituent republic of the USSR.
1944 Some 160,000 Meskhetian Turks forcibly transported from their

native Georgia to Uzbekistan by Stalin.

1989 June: Tashlak, Yaipan, and Ferghana were the scenes of riots in which Meskhetian Turks were attacked; 70 killed and 850 wounded.

1990 June: economic and political sovereignty declared; Islam Karimov became president.

1991 March: Uzbek supported 'renewed federation' in USSR referendum. Aug: anti-Gorbachev coup in Moscow initially accepted by President Karimov; later, Karimov resigned from Soviet Communist Party (CPSU) Politburo; Uzbek Communist Party (UCP) broke with CPSU; pro-democracy rallies dispersed by militia; independence declared. Dec: joined new Commonwealth of Independent States (CIS).

1992 Jan: admitted into Conference on Security and Cooperation in Europe (CSCE); violent food riots in Tashkent. March: joined the United Nations (UN); US diplomatic recognition achieved.

Vanuatu

Republic of
(*Ripablik Blong
Vanuatu*)

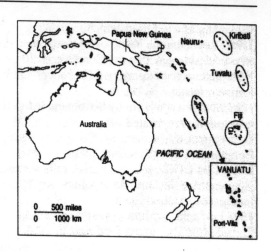

area 14,800 sq km/5,714 sq mi

capital Vila, or Port-Vila, (on Efate)

towns Luganville (on Espíritu Santo)

physical comprises around 70 islands, including Espíritu Santo, Malekula, and Efate; densely forested, mountainous

head of state Fred Timakata from 1989

head of government Maxime Carlot from 1991

political system democratic republic

exports copra, fish, coffee, cocoa

currency vatu

population (1989) 152,000 (90% Melanesian); growth rate 3.3% p.a.

life expectancy men 67, women 71 (1989)

languages Bislama 82%, English, French (all official)

literacy 53%

religions Presbyterian 40%, Roman Catholic 16%, Anglican 14%, animist 15%

GDP $125 million (1987); $927 per head

chronology

1906 Islands jointly administered by France and Britain.

1975 Representative assembly established.

1978 Government of national unity formed, with Father Gerard

Leymang as chief minister.

1980 Revolt on the island of Espíritu Santo delayed independence but it was achieved within the Commonwealth, with George Kalkoa (adopted name Sokomanu) as president and Father Walter Lini as prime minister.

1988 Dismissal of Lini by Sokomanu led to Sokomanu's arrest for treason. Lini reinstated.

1989 Sokomanu sentenced to six years' imprisonment; succeeded as president by Fred Timakata.

1991 Lini voted out by party members; replaced by Donald Kalpokas. General election produced Union of Moderate Parties–Vanuatu National Party coalition under Maxime Carlot.

Vatican City State

(*Stato della Città del Vaticano*)

area 0.4 sq km/109 acres

physical forms an enclave in the heart of Rome, Italy

head of state and government John Paul II from 1978

political system absolute Catholicism

currency Vatican City lira; Italian lira

population (1985) 1,000

languages Latin (official), Italian

religion Roman Catholic

chronology

1929 Lateran Treaty recognized sovereignty of the pope.

1947 New Italian constitution confirmed the sovereignty of the Vatican City State.

1978 John Paul II became the first non-Italian pope for more than 400 years.

1985 New concordat signed under which Roman Catholicism ceased to be Italy's state religion.

Venezuela
Republic of
(*República de
Venezuela*)

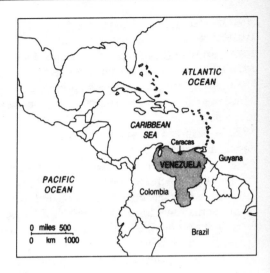

area 912,100 sq km/352,162 sq mi
capital Caracas
towns Barquisimeto, Valencia, Maracaibo
physical Andes Mountains and Lake Maracaibo in NW; central plains
(llanos); delta of river Orinoco in E; Guiana Highlands in SE
head of state and of government Carlos Andrés Pérez from 1988
government federal democratic republic
exports coffee, timber, oil, aluminium, iron ore, petrochemicals
currency bolívar
population (1990 est) 19,753,000 (mestizos 70%, white (Spanish,
Portuguese, Italian) 20%, black 9%, Amerindian 2%); growth rate
2.8% p.a.
life expectancy men 67, women 73 (1989)
religions Roman Catholic 96%, Protestant 2%
languages Spanish (official), Amerindian languages 2%
literacy 88% (1989)
GNP $47.3 bn (1988); $2,629 per head (1985)
recent chronology
1961 New constitution adopted, with Rómulo Betancourt as president.

1964 Dr Raúl Leoni became president.

1969 Dr Rafael Caldera became president.

1974 Carlos Andrés Pérez became president.

1979 Dr Luis Herrera became president.

1984 Dr Jaime Lusinchi became president; social pact established between government, trade unions, and business; national debt rescheduled.

1987 Widespread social unrest triggered by inflation; student demonstrators shot by police.

1988 Carlos Andrés Pérez elected president. Payments suspended on foreign debts (increase due to drop in oil prices).

1989 Economic austerity programme enforced by $4.3 billion loan from International Monetary Fund (IMF). Price increases triggered riots; 300 people killed. Feb: martial law declared. May: General strike. Elections boycotted by opposition groups.

1991 Protests against austerity programme continued.

1992 Attempted anti-government coups failed. Pérez promised constitutional changes.

Vietnam
Socialist Republic of
(*Công Hòa Xã Hôi
Chu Nghĩa Viêt
Nam*)

area 329,600 sq km/127,259 sq mi
capital Hanoi
towns ports Ho Chi Minh City (formerly Saigon), Da Nang, Haiphong
physical Red River and Mekong deltas, centre of cultivation and
population; tropical rainforest; mountainous in N and NW
environment during the Vietnam War an estimated 2.2 million
hectares/5.4 million acres of forest were destroyed. The country's
National Conservation Strategy is trying to replant 500 million trees
each year
head of state Le Duc Anh from 1992
head of government Vo Van Kiet from 1991
political system communism
exports rice, rubber, coal, iron, apatite
currency dong
population (1990 est) 68,488,000 (750,000 refugees, mainly ethnic
Chinese, left 1975–79; some settled in China, others fled by sea—the
'boat people'—to Hong Kong and elsewhere); growth rate 2.4% p.a.

life expectancy men 62, women 66 (1989)
languages Vietnamese (official), French, English, Khmer, Chinese, local languages
religions Buddhist, Taoist, Confucian, Christian
literacy 78% (1989)
GNP $12.6 bn; $180 per head (1987)
chronology
1945 Japanese removed from Vietnam at end of World War II.
1946 Commencement of Vietminh war against French.
1954 France defeated at Dien Bien Phu. Vietnam divided along 17th parallel.
1964 US troops entered Vietnam War.
1973 Paris cease-fire agreement.
1975 Saigon captured by North Vietnam.
1976 Socialist Republic of Vietnam proclaimed.
1978 Admission into Comecon. Vietnamese invasion of Cambodia.
1979 Sino-Vietnamese border war.
1986 Retirement of 'old guard' leaders.
1987–88 Over 10,000 political prisoners released.
1988–89 Troop withdrawals from Cambodia continued.
1989 'Boat people' leaving Vietnam murdered and robbed at sea by Thai pirates. Troop withdrawal from Cambodia completed. Hong Kong forcibly repatriated some Vietnamese refugees.
1991 Vo Van Kiet replaced Do Muoi as prime minister. Cambodia peace agreement signed. Relations with China normalized.
1992 Sept: Le Duc Anh elected president. Dec: relations with South Korea normalized; USA eased 30-year-old trade embargo.

Yemen
Republic of
(*al Jamhuriya al
Yamaniya*)

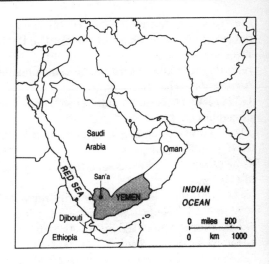

area 531,900 sq km/205,367 sq mi
capital San'a
towns Ta'iz, Aden
physical hot moist coastal plain, rising to plateau and desert
head of state and of government Ali Abdullah Saleh from 1990
political system authoritarian republic
exports cotton, coffee, grapes, vegetables
currency rial
population (1990 est) 11,000,000; growth rate 2.7% p.a.
life expectancy men 47, women 50
language Arabic
religions Sunni Muslim 63%, Shi'ite Muslim 37%
literacy men 20%, women 3% (1985 est)
GNP $4.9 bn (1983); $520 per head
chronology
1918 Yemen became independent.
1962 North Yemen declared the Yemen Arab Republic (YAR), with
Abdullah al-Sallal as president. Civil war broke out between royalists
and republicans.

1967 Civil war ended with the republicans victorious. Sallal deposed and replaced by Republican Council. The People's Republic of South Yemen was formed.

1970 People's Republic of South Yemen renamed People's Democratic Republic of Yemen.

1971–72 War between South Yemen and the YAR; union agreement signed but not kept.

1974 Ibrahim al-Hamadi seized power in North Yemen and Military Command Council set up.

1977 Hamadi assassinated and replaced by Ahmed ibn Hussein al-Ghashmi.

1978 Constituent People's Assembly appointed in North Yemen and Military Command Council dissolved. Ghashmi killed by envoy from South Yemen; succeeded by Ali Abdullah Saleh. War broke out again between the two Yemens. South Yemen president deposed and Yemen Socialist Party (YSP) formed with Abdul Fattah Ismail as secretary general, later succeeded by Ali Nasser Muhammad.

1979 Cease-fire agreed with commitment to future union.

1983 Saleh elected president of North Yemen for a further five-year term.

1984 Joint committee on foreign policy for the two Yemens met in Aden.

1985 Ali Nasser re-elected secretary general of the YSP in South Yemen; removed his opponents. Three bureau members were killed.

1986 Civil war in South Yemen; Ali Nasser dismissed. New administration under Haydar Abu Bakr al-Attas.

1988 President Saleh re-elected in North Yemen.

1989 Draft constitution for single Yemen state published.

1990 Border between two Yemens opened; countries formally united 22 May as Republic of Yemen.

1991 New constitution approved.

1992 Anti-government riots.

1993 Multiparty parliamentary elections set for April.

Yugoslavia

area 88,400 sq km/34,100 sq mi
capital Belgrade
towns Kraljevo, Leskovac, Pristina, Novi Sad, Titograd
physical mountainous, with river Danube plains in N and E; limestone
(Karst) features in NW
head of state Dobrica Cosic from 1992
head of government Radoje Kontic from 1993
political system socialist pluralist republic
exports machinery, electrical goods, chemicals, clothing, tobacco
currency dinar
population (1990) 12,420,000 (Serb 53%, Albanian 15%, Macedonian
11%, Montenegrin 5%,Muslim 3%,Croat 2%)
life expectancy men 69, women 75 (1989)
languages Serbian variant of Serbo-Croatian, Macedonian, Slovenian
religion Eastern Orthodox 41% (Serbs), Roman Catholic 12%
(Croats), Muslim 3%
literacy 90% (1989)
GNP $154.1 bn; $6,540 per head (1988)

chronology

1918 Creation of Kingdom of the Serbs, Croats, and Slovenes.

1929 Name of Yugoslavia adopted.

1941 Invaded by Germany.

1945 Yugoslav Federal Republic formed under leadership of Tito; communist constitution introduced.

1948 Split with USSR.

1961 Nonaligned movement formed under Yugoslavia's leadership.

1980 Tito died; collective leadership assumed power.

1988 Economic difficulties: 1,800 strikes, 250% inflation, 20% unemployment. Ethnic unrest in Montenegro and Vojvodina; party reshuffled and government resigned.

1989 Reformist Croatian Ante Marković became prime minister. 29 died in ethnic riots in Kosovo province, protesting against Serbian attempt to end autonomous status of Kosovo and Vojvodina; state of emergency imposed. May: inflation rose to 490%; tensions with ethnic Albanians rose.

1990 Multiparty systems established in Serbia and Croatia.

1991 June: Slovenia and Croatia declared independence, resulting in clashes between federal and republican armies; Slovenia accepted EC-sponsored peace pact. Fighting continued in Croatia; repeated calls for cease-fires failed. Dec: President Stipe Mesic and Prime Minister Ante Marković resigned.

1992 Jan: EC-brokered cease-fire established in Croatia; EC and USA recognized Slovenia's and Croatia's independence. Bosnia-Herzegovina and Macedonia declared independence. April: Bosnia-Herzegovina recognized as independent by EC and USA amid increasing ethnic hostility; bloody civil war ensued. New Federal Republic of Yugoslavia (FRY) proclaimed by Serbia and Montenegro but not recognized externally. May: Western ambassadors left Belgrade. International sanctions imposed against Serbia and Montenegro. Hostilities continued. Jun: Dobrica Cosic became president. Jul: Milan Panic became prime minister. Sept: UN membership suspended. Dec: Slobodan Milosevic re-elected Serbian president; Panic removed from office in vote of no confidence.

1993 Radoje Kontic became prime minister.

Zaire
Republic of
(*République
du Zaíre*)
(formerly *Congo*)

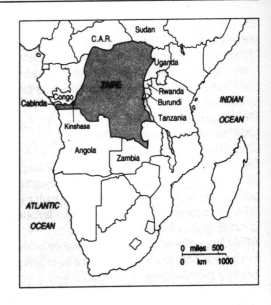

area 2,344,900 sq km/905,366 sq mi
capital Kinshasa
towns Lubumbashi, Kananga, Kisangani, Matadi, Boma
physical Zaïre River basin has tropical rainforest and savanna;
mountains in E and W
head of state Mobutu Sese Seko Kuku Ngbendu wa Zabanga
head of government Faustin Birindwa from 1993
political system socialist pluralist republic
exports coffee, copper, cobalt (80% of world output), industrial
diamonds, palm oil
currency zaïre
population (1990 est) 35,330,000; growth rate 2.9% p.a.
life expectancy men 51, women 54 (1989)
languages French (official), Swahili, Lingala, other African
languages; over 300 dialects
religions Christian 70%, Muslim 10%
literacy men 79%, women 45% (1985 est)

GNP $5 bn (1987); $127 per head
chronology
1908 Congo Free State annexed to Belgium.
1960 Independence achieved from Belgium as Republic of the Congo. Civil war broke out between central government and Katanga province.
1963 Katanga war ended.
1967 New constitution adopted.
1970 Col Mobutu elected president.
1971 Country became the Republic of Zaire.
1972 The Popular Movement of the Revolution (MPR) became the only legal political party. Katanga province renamed Shaba.
1974 Foreign-owned businesses and plantations seized by Mobutu and given in political patronage.
1977 Original owners of confiscated properties invited back. Mobutu re-elected; Zairians invaded Shaba province from Angola, repulsed by Belgian paratroopers.
1978 Second unsuccessful invasion from Angola.
1988 Potential rift with Belgium avoided.
1990 Mobutu announced end of ban on multiparty politics, following internal dissent.
1991 Multiparty elections promised. Sept: after antigovernment riots, Mobutu agreed to share power with opposition; Etienne Tshisekedi appointed premier. Oct: Tshisekedi dismissed.
1992 Aug: Tshisekedi reinstated against Mobuto's wishes; interim opposition parliament formed. Oct: renewed rioting.
1993 Jan: French ambassador shot dead by loyalist troops during army mutiny; France and Belgium prepared to evacuate civilians. March: Tshisekedi dismissed by Mobutu, replaced by Faustin Birindwa, but Tshisekedi still considered himself in office.

Zambia
Republic of

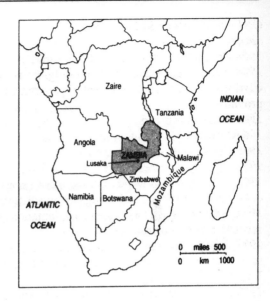

area 752,600 sq km/290,579 sq mi
capital Lusaka
towns Kitwe, Ndola, Kabwe, Chipata, Livingstone
physical forested plateau cut through by rivers
head of state and government Frederick Chiluba from 1991
political system socialist pluralist republic
exports copper, cobalt, zinc, emeralds, tobacco
currency kwacha
population (1990 est) 8,119,000; growth rate 3.3% p.a.
life expectancy men 54, women 57 (1989)
language English (official); Bantu dialects
religions Christian 66%, animist, Hindu, Muslim
literacy 54% (1988)
GNP $2.1 bn (1987); $304 per head (1986)
chronology
1899–1924 As Northern Rhodesia, under administration of the British
South Africa Company.

1924 Became a British protectorate.

1964 Independence achieved from Britain, within the Commonwealth, as the Republic of Zambia with Kenneth Kaunda as president.

1972 United National Independence Party (UNIP) declared the only legal party.

1976 Support for the Patriotic Front in Rhodesia declared.

1980 Unsuccessful coup against President Kaunda.

1985 Kaunda elected chair of the African Front Line States.

1987 Kaunda elected chair of the Organization of African Unity (OAU).

1988 Kaunda re-elected unopposed for sixth term.

1990 Multiparty system announced for 1991.

1991 Movement for Multiparty Democracy won landslide election victory; Frederick Chiluba became president.

1992 Food and water shortages caused by severe drought.

Zimbabwe
Republic of

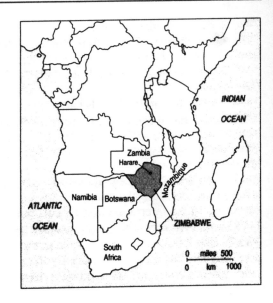

area 390,300 sq km/150,695 sq mi
capital Harare
towns Bulawayo, Gweru, Kwekwe, Mutare, Hwange
physical high plateau with central high veld and mountains in E; rivers
Zambezi, Limpopo
head of state and government Robert Mugabe from 1987
political system effectively one-party socialist republic
exports tobacco, asbestos, cotton, coffee, gold, silver, nickel,
copper
currency Zimbabwe dollar
population (1990 est) 10,205,000 (Shona 80%, Ndbele 19%; about
100,000 whites); growth rate 3.5% p.a.
life expectancy men 59, women 63 (1989)
languages English (official), Shona, Sindebele
religions Christian, Muslim, Hindu, animist
literacy men 81%, women 67% (1985 est)
GNP $5.5 bn (1988); $275 per head (1986)

chronology

1889–1923 As Southern Rhodesia, under administration of British South Africa Company.

1923 Became a self-governing British colony.

1961 Zimbabwe African People's Union (ZAPU) formed, with Joshua Nkomo as leader; declared illegal 1962.

1963 Zimbabwe African National Union (ZANU) formed, with Robert Mugabe as secretary general.

1964 Ian Smith became prime minister. ZANU banned. Nkomo and Mugabe imprisoned.

1965 Smith declared unilateral independence.

1974 Nkomo and Mugabe released.

1975 Geneva conference set date for constitutional independence.

1979 Smith produced new constitution and established a government with Bishop Abel Muzorewa as prime minister. New government denounced by Nkomo and Mugabe. Conference in London agreed independence arrangements (Lancaster House Agreement).

1980 Independence achieved from Britain, with Robert Mugabe as prime minister and Rev Canaan Banana as president.

1982 Nkomo dismissed from the cabinet, leaving the country temporarily.

1984 ZANU–Patriotic Front (PF) party congress agreed to create a one-party state in future.

1985 Relations between Mugabe and Nkomo improved. Troops sent to Matabeleland to suppress rumoured insurrection.

1987 Separate assembly seats for whites abolished. President Banana retired; Mugabe combined posts of head of state and prime minister and took on the title executive president.

1988 Nkomo returned to the cabinet and was appointed vice president.

1989 Opposition party, the Zimbabwe Unity Movement, formed by Edgar Tekere; draft constitution drawn up, renouncing Marxism–Leninism; ZANU–PF and ZAPU formally merged.

1990 ZANU–PF re-elected. State of emergency ended. Opposition to creation of one-party state.

1992 United Front formed to oppose ZANU–PF. March: Mugabe declared dire drought and famine situation a national disaster